To the memory of my parents

Contents

This is not a book about the future of technology. That is just as well: history is littered with examples of people who guess wrongly. Either they expect great things of a technology that proves to be a blind alley – or, more often, they fail to spot the successes. The personal computer, the home fax, the mobile phone, the Internet itself: all were largely unnoticed by pundits until consumers grabbed them.

Besides, what matters most about a new technology is not how it works, but how people use it, and the changes it brings about in human lives. In the twentieth century, lives were changed beyond measure by the spread of electricity. In the century ahead, the communications revolution may be no less important, although in quite different ways.

In fact, this book is about a group of technologies that already exist. They will be constantly refined and improved in the coming years, just as the design of the car or the aircraft was refined all through the last century. But their broad shape is already clear enough to allow speculation on the impacts they are likely to have. The key ideas are summarised in "The Trendspotter's Guide" that appears right after this preface. The rest of the book develops these core ideas.

The most important single theme is the economic and social importance of the steep fall in the distance premium for communications. The cost of communicating ideas and data is now distance-free, thanks to the Internet and to the dramatic change in telephone tariffs. This astonishing change, which has occurred in less than a decade, has come at a time when other forces such as globalisation and international migration are also changing the ways people think about distance.

. . . . .

But the Death of Distance (a phrase drawn from *The Tyranny of Distance*, Geoffrey Blainey's classic study of the impact of isolation on Australia) is of course by no means the only impact of the communications revolution. Its effects are enhanced by the sheer versatility of the Internet. It is a library, a marketplace, a meeting place, a computing platform, a distribution chain. Its amazing ability to link together many different activities is improving all the time. It is also extraordinarily accessible, and will become even more so, as more and more devices allow people to use the Internet without ever realising that they are doing so.

Side-by-side with the Internet, the sweep of the mobile telephone across the globe is bringing sophisticated communications even to the world's poorer countries. A simple innovation – the prepaid mobile-phone card – is making the telephone affordable to millions of people who would never risk the uncertain expense of a monthly bill, or who would have had to wait years for a connection. At last, the telephone, with all its array of possible services, is becoming universally available around the world.

Some of the ideas that follow have had earlier airings, often in the columns of *The Economist*. As the endnotes show, I have pillaged my own and my colleagues' writings. This British magazine, based in London, is an ideal vantage point from which to watch the distance-free world evolve: the readers are scattered around the globe, with two-thirds of the circulation in North America and most of the rest outside Britain. The perspective of the book is similar to that of *The Economist*. While most analysis of the communications revolution has been from an American viewpoint, reflecting American dominance of the Internet, this book seeks to be truly international.

The book is a largely rewritten version of a work with the same name, published in 1997. It began as a simple update of the earlier edition; it soon became clear that so much of what had been prediction was now fact that a completely new text was needed. More than half the material in this edition is therefore new, including the chapter on history, which aims to draw lessons from earlier advances in communications. Entirely new, too, are the chapters on commerce. In particular, the transforming effects of the communications revolution on the internal workings of companies have become even more apparent in the intervening three years.

Since the original book appeared, excitement about the revolutionary impact of new communications has mounted to a crescendo – and then died away even more abruptly, with the collapse of the market for technology stocks. Both responses are equally wrong. The most important impacts will take at least a decade to fall into place. Indeed, there will be changes in the way public services are organized and delivered that will take a full generation to come through. Because profound change happens gradually, people may not really notice it until long after it has occurred.

The original edition was built on the advice and ideas of four people: Chris Anderson and Azeem Azhar, my colleagues at *The Economist*; Tim Kelly; and my husband, Hamish McRae. This edition has drawn on the ideas and writings of Andrew Odlyzko, whose unpublished work on the importance of connectivity rather than content has been an important influence. Lewis Neal read the whole book, checked facts and came up with invaluable new information. Once again, though, my biggest debt by far is to my husband. Many of the best points in the book are his ideas. Anybody who has read his columns in *The Independent* newspaper or his own book, *The World in 2020*, will see a continuity of thought. This book is really his as much as mine.

London, December 2000

The Trendspotter's Guide to New Communications

How will the death of distance shape the future? Here are some of the most important developments to watch, each discussed in depth later in this book.

1. The Death of Distance. Distance will no longer decide the cost of communicating electronically. Indeed, once investment has been made in a communications network, in buying a computer or telephone, or in setting up a Web site, the additional cost of sending or receiving an extra piece of information will be virtually zero.

2. The Fate of Location. Companies will be free to locate many screen-based activities wherever they can find the best bargain of skills and productivity. Developing countries will increasingly perform on-line services – including monitoring security screens, inputting data from forms, running help-lines, and writing software code – and sell them to the rich industrial countries that generally produce such services domestically.

3. Improved Connections. Most people on earth will eventually have access to networks that are all interactive and broadband. The Internet will continue to exist in its present form, but will also carry many other services, including telephone and television.

4. Increased Mobility. Every form of communication will be available for mobile or remote use.

5. More Customized Networks. The huge capacity of networks will enable individuals to order "content for one": that is, individual consumers will receive (or send) exactly what they want to receive (or send), when and where they want it.

6. A Deluge of Information. Because people's capacity to absorb new information will not increase, they will need filters to sift, process, and edit it.

7. Increased Value of Brand. Companies will want ways to push their information ahead of their competitors'. One of the most effective will be branding. What's hot – whether a product, a personality, a sporting event, or the latest financial data – will attract the greatest rewards.

8. More Minnows, More Giants. Many of the costs of starting a new business will fall and companies will more easily buy in services. So small companies will start up more readily, offering services that, in the past, only giants had the scale and scope to provide. If they can back creativity with competence and speed, they will compete effectively with larger firms. At the same time, communication amplifies the strength of brands and the power of networks. In industries where networks matter, concentration will increase.

9. More Competition. More companies and customers will have access to accurate price information. In addition, some entry barriers will fall. The result will be greater competition in many markets, resulting in "profitless prosperity": it will be easier to find buyers, but harder to make fat margins.

10. Increased Value of Niches. The power of the computer to search, identify, and classify people according to similar needs and tastes will create sustainable markets for many niche products. One of the most valuable improvements will be in the ability of people to locate things that have hitherto been hard to find: from friends with similar tastes to specialized services.

11. Communities of Practice. The horizontal bonds among people performing the same job or speaking the same language in different parts of the world will strengthen. Common interests, experiences, and pursuits, rather than proximity, will bind these communities together.

12. The Loose-Knit Corporation. Culture and communications networks, rather than rigid management structures, will hold companies together. Vertically integrated companies that do everything from buying the raw materials to repairing their own products will disappear. Internet-based technologies will reduce

the costs of dealing with arm's-length suppliers and partners. Alliances will bond companies together at many levels.

13. Openness as a Strategy. Loyalty, trust, and open communications will reshape the nature of supplier and customer contacts. Suppliers will draw directly on their customers' databases, working as closely and seamlessly as an in-house supplier does now. Customers will be able to manage and track their orders through the production process.

14. Manufacturers as Service Providers. Companies will tailor their products more precisely to a customer's tastes and needs. Some will retain lasting links with their products: car companies, for instance, will continue electronically to track, monitor, and learn about their vehicles throughout the product's life cycle. New opportunities to build links with customers will emerge as a result.

15. The Inversion of Home and Office. The line between home and work will blur. People will increasingly work from home and shop from work. The office will become a place for the social aspects of work such as networking, brainstorming, lunching, and gossiping. More people will work on the move: from their cars, from hotel rooms, from airport departure lounges. Home design will change: new homes will routinely have home offices.

16. The Proliferation of Ideas. New ideas and information will travel faster to the remotest corners of the world. Developing countries will acquire more rapidly access to the industrial world's knowledge and ideas. That will help many developing countries to grow more quickly and even to narrow the gap with the rich world.

17. The Decline of National Authority. Governments will find national legislation and censorship inadequate for regulating the global flow of information. As content sweeps across national borders, it will be harder to enforce laws banning child pornography, libel, and other criminal or subversive material, and those protecting copyright and other intellectual property.

18. Loss of Privacy. Protecting privacy will be difficult, as it was in the villages of past centuries. Governments and companies will easily monitor people's movements. Machines will recognize physical attributes such as a voice or fingerprint. Civil

libertarians will worry, but others will rationalize the loss as a fair exchange for the reduction of crime, including fraud and illegal immigration. In the electronic village, there will be little true privacy – and little unsolved crime.

19. A Global Premium for Skills. Pay differentials will continue to widen, as companies fight for the scarce talents of well educated workers. Managerial and professional jobs will be less vulnerable to competition from automation than jobs requiring relatively little skill. In addition, the Internet enhances the value of creative use of information. On-line recruitment will make the job market more global and efficient. As a result, highly skilled people will earn broadly similar amounts, wherever they live in the world.

20. Rebirth of Cities. As individuals spend less time in the office and more time working from home or on the road, cities will change from concentrations of office employment to centers of entertainment and culture. They will become places where people congregate to visit museums and galleries, attend live performances of all kinds, participate in civic events, and dine in good restaurants. Some poor countries will use low-cost communications to stem the flight from the countryside by providing rural areas with better medical services, jobs, education, and entertainment.

21. The Rise of English. The global role of English as a second language will continue. It will become the global communications standard: the default language of the electronic world.

22. Communities of Culture. At the same time, electronic communications will reinforce less widespread languages and cultures, not replace them with Anglo-Saxon and Hollywood. The falling cost of creating and distributing many entertainment products will also reinforce local cultures and help scattered peoples and families to preserve their cultural heritage.

23. A New Trust. Since it will be easier to check whether people and companies deliver what they have promised, many services will become more reliable and people will be more likely to trust each other to keep their word. However, those who fail to deliver will quickly lose that trust, which will be increasingly hard to regain.

24. People as the Ultimate Scarce Resource. The key challenge for companies will be to hire and retain good people, motivating them while at the same time extracting value from them. A company will constantly need to convince its best employees that working for it enhances their value as well as its own.

25. Global Peace. Democracy will continue to spread: people who live under dictatorial regimes will be more aware of their governments' failures. Democracies have always been more reluctant to fight than dictatorships. In addition, countries will grow yet more economically interdependent. People will communicate more freely with human beings on other parts of the globe. As a result, while wars will still be fought, the effect may be to foster world peace.

Now read on.

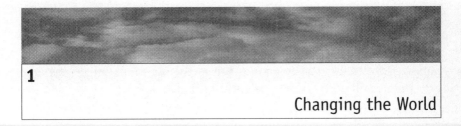

To understand where we are today, imagine the automobile in 1910. Twenty-five years after its invention, it had already evolved into a form we would recognize now. It had an engine at the front, pneumatic tyres, a clutch, and a gearbox, and its most advanced models could travel at present-day speeds. Most important of all, Henry Ford had started to mass-produce it, setting it on the way to becoming a standard consumer item.

Yet the immense social consequences of the automobile's development took most of the twentieth century to unfold. Highways gradually paved the routes from railway stations and city centers to suburbs and supermarkets. Jobs disappeared (who these days knows what an ostler did?), even as millions more were created. The automobile built new industries and created markets for new skills (truck-driving, vehicle repairs). It made possible, particularly in North America, the flight of the middle classes from the inner city. And it liberated ordinary people (including women) to travel where and when they wished.

This book starts from the assumption that technology, driving economics, has the power to change the social and physical world. The technologies of the communications revolution have advanced at least as far as had those of the automobile in 1910. Their outlines are clear: voice, video, and data, digitally delivered across interconnecting networks to a variety of different terminals with different uses. As always, the technology takes shape long before its full consequences for society emerge.

Still hazy are the long-term effects: the equivalent in this century of the social and economic changes wrought by the automobile in the last. One thing, though, is certain. The death of distance and the communications revolution will be among the most important forces shaping economies and society in the next fifty years or so.

Think of it as one of three great revolutions in the cost of transport. The nineteenth century, dominated by the steamship and the railway, saw a transformation in the cost of transporting goods; the twentieth century, with first the motor car and then the aeroplane, in the cost of transporting people. The new century will be dominated by the transformation in the cost of transporting knowledge and ideas. That revolution began long ago with the mail, proceeded through the telegraph, the telephone, and broadcasting, and has culminated in a group of innovations that are racing forward with amazing speed.

The death of distance is only one manifestation of the astonishing changes taking place as communications and computers are combined in new ways. High-capacity fiber-optic networks and digital compression already carry voice, video, and data around the globe so efficiently that the additional cost of sending a message an extra hundred miles is effectively zero. The Internet, an invention that only began to be accessible to ordinary folk in the 1990s, has introduced perhaps 385 million people around the world to the idea that it costs no more to visit a book store in Seattle than one in the local high street.[1]

Wireless – the other communications revolution – is simultaneously killing location, putting the world in our pockets.[2] Indeed, globally, more people have mobile telephones than Internet connections. [Fig 1-1]

That these technologies will change the world is beyond a doubt. The way they will do so is more mysterious. Big changes in distribution costs can have enormous effects. They tend to be liberating and enriching. They are also invariably democratizing. Thus the decline in the costs of ocean freight revolutionized the human diet, allowing ordinary people in the industrial world to enjoy a variety of food that could once be afforded only by the wealthy. The even more dramatic decline in airtransport costs has enabled ordinary tourists to visit places that were once the preserve of the rich.

It can take many years for such technology-driven transitions to occur. The change in diet took half a century; the rise of mass tourism, now the world's largest industry, took about fifteen years. The fall in

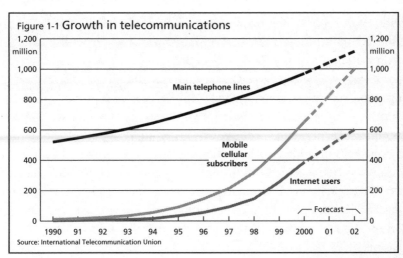

Figure 1-1 **Growth in telecommunications**

Source: International Telecommunication Union

the cost of a telephone call or a satellite link has been even more substantial, and it is happening faster. But we are only beginning to see the effects. Most shoppers around the world are not on the Internet; most businesses do not sell on-line; most Web sites make no money. Even in the United States, which in early 2000 accounted for 90 percent of the world's Web sites and three-quarters of e-commerce revenue, consumer purchases on-line amounted to only 0.7 percent of all retail sales in the first four months of the year.[3] For all the millions of Web sites created, 83 percent of Internet users go to no more than ten domain names each week.[4]

Innovation is a curious business. It rarely happens instantaneously. Technologies develop for one purpose and then – as the Internet's history shows – turn out to be more useful for another. Much innovation is a matter of small refinements over a long period, turning an unreliable novelty into a device with an evolving range of applications. "The full importance of an epoch-making idea is often not perceived in the generation in which it is made," observed Alfred Marshall, one of the fathers of modern economics, in his *Principles of Economics*. "A new discovery is seldom fully effective for practical purposes till many minor improvements and subsidiary discoveries have gathered themselves around it."[5]

The innovations taking place in electronic communications will be far more pervasive than some of the advances with which they are often compared, such as the railways or the telegraph. The Internet, in particular, is remarkable for its many different capabilities. It is more than just a means of communicating: it is a marketplace, a library, a distribution chain. It connects individuals, companies, machines. Indeed, it connects most of the world's computers, to form a vast computer in its own right. The novelties that it allows will be refined and adapted more quickly than has happened with earlier inventions. One reason for this is that people are more aware than they were during any past era of large technological change of the commercial potential of what is happening, and indeed of the way the process of innovation tends to work. Another reason is that communications technologies have an intrinsic capacity to disseminate new ideas. They also go with the flow. Their border-hopping quality will reinforce and exploit the changes brought about by globalization and the introduction of the euro, for instance.

Some of the changes caused by the death of distance appear quickly. Instant messaging has created new kinds of on-line socializing, while the short-message system on mobile telephones has transformed the night life of the young in Europe and Asia. Other changes will take a generation or two to emerge. Although the communications revolution may already seem an old story when viewed from Palo Alto, its main effects will come as it spreads around the world and becomes embedded in new investments and organizations.

Before that happens, many obstacles must be overcome. Take two. Fast Internet access for small businesses (and households) requires a large investment in connections between "backbone" cable and the neighborhood. For the Internet to become as useful in Asia, where most of the world's people live, as it is in North America, companies need to offer Internet addresses in alphabets other than Roman characters for the benefit of the Chinese, Japanese, Korean, and other Asian users to whom the letters ".com" mean little. That process began only in 2000.[6]

As such changes go forward, they will alter, in ways that are only dimly imaginable, decisions about where people work and what kind of work they do, concepts of national borders and sovereignty, and patterns of international trade.

Some of the revolution's most powerful effects will be on the provision of services. Indeed, it has the potential to raise the productivity and quality of service provision in the way mass production raised the efficiency and quality of manufacturing. As Brad DeLong, an economist at the University of California at Berkeley, puts it, "Information technology and the Internet amplify brain power in the same way that the technologies of the industrial revolution amplified muscle power."[7]

Brain power is the essential ingredient of many services, and the largest provider of services in most countries is the state. Government is one of the slowest organizations in society to change, so the full impact here will take a generation or two to emerge. But few things make more difference to the quality of people's lives than the efficient provision of public services, from policing to education to traffic management to health care in old age.

The death of distance loosens the grip of geography. It does not destroy it. A drive along Silicon Valley's Sandhill Road, epicenter of the Internet revolution, or a visit to a meeting of First Tuesday, an organization that holds social events in cities around the world to introduce potential dotcom investors to would-be clients, is immediate proof of the enduring importance of proximity in commercial and social life. Yet vast new opportunities are appearing. Companies will have more freedom to locate a service where their key staff want to live, rather than near its market. People – some at least – will gain more freedom to live far from their employers. Some kinds of work will be organized in three shifts, to maximize the use of the world's three main time zones: the Americas, East Asia/Australia, and Europe. Time zones and language groups, rather than mileage, will come to define distance.

Suppose we are on the verge of a new era, when all sorts of activities that we once thought of as local – health care, education, financial and legal advice – become available globally. One consequence would be a replication of the democratizing effect of the falling costs of ocean freight and air transport. Services that could once be afforded only by the wealthy might, at last, be available to the many. Ordinary people would be, once again, liberated and enriched.

Barriers and borders will break down. The horizontal bonds among people with the same interest, or speaking the same language in different parts of the world, will grow stronger. Some of society's vertical bonds – between government and the governed, or bosses and workers – will grow weaker. Undoubtedly, governments will face problems: their jurisdictions will no longer encompass the forces that affect their citizens' lives. As for companies, they will become looser structures held together mainly by their cultures and their communications networks, yet at the same time they will become more closely integrated with other firms with which they do business. For individuals, the lines between work and leisure will grow less distinct. The design of the office and of the home will alter to accommodate the changing patterns of this communications-driven life.

New ideas will spread faster, leaping borders. Entire electorates will learn things that once only a few bureaucrats knew. Small companies will offer services that previously only giants could provide. Poor countries will have immediate access to information that was once restricted to the industrial world and travelled only slowly, if at all, beyond it. In all of these ways, the communications revolution is profoundly democratic and liberating, levelling the imbalance between large and small, rich and poor. The death of distance, overall, should be welcomed and enjoyed.

The Road Ahead

· · · · ·

This book attempts to guess where all these changes will lead. It looks out beyond the turmoil in financial markets which fashion for, and then disillusionment with, high-technology stocks has caused. As so often happens with new technologies, the immediate possibilities tend to be exaggerated, the long-term possibilities underestimated.

The book begins with a dose of history, both to put what is happening in context and to see what lessons the past can offer for the future. Three points stand out with singular clarity. One – particularly humbling for anybody writing about the future – is how frequently projections are wrong. Ever since, in 1816, Britain's Admiralty brushed aside an early inventor of the electronic telegraph with the firm assertion that "Telegraphs of any kind are wholly unnecessary," people have been misjudging the technological future (and thus its consequences).[8]

The second point is the far-reaching effects that earlier advances in communications have had. The third, a theme discussed elsewhere in the book, is the importance of getting pricing structures right. When the Victorian postal system in Britain applied death-of-distance pricing to the domestic mail, the volume of mail doubled in two years.

The book then turns to the three main technologies at the heart of the revolution – voice, video, and data – to look at how their development weaves together. In exploring the future, it helps to begin with the present. "And in today, already walks tomorrow," as Samuel Taylor Coleridge more poetically put it. So understanding the relative strengths of the two main existing pillars of communications, the telephone and the television, is important in gauging how tomorrow's digital networks may develop.

Here, convergence is an important issue. Because it is possible to watch television on a home computer connected to a telephone line, many people have assumed that this is what folk will want to do. Indeed, some of the largest mergers in history – AT&T with TCI, say, or America Online with Time Warner – were based at least partly on the assumption that the roles of the telephone, television, and PC would increasingly overlap. Now it is the turn of the mobile telephone: because it can give access to the Internet, enormous investment is going into this dream convergence, creating products that offer that possibility.

Yet consumers have the power to surprise. Innovations that are relatively primitive in technological terms may turn out to be those that meet people's needs better than more sophisticated novelties. Thus e-mail has proved to be the killer application of the Internet; and the short-message system, which uses the mobile telephone as a pager, has become an unexpected hit across Europe. It may be that convergence is not at all what consumers want – or not in the forms that have so far been devised. Certainly connectivity is likely to prove more appealing than content.

In addition, the way people adopt new technologies will vary from country to country. That has always been true. Stockholm at the start of the twentieth century had more telephones than the whole of the United Kingdom.[9] In spring 2000, 76 percent of Swedes had mobile telephones, the highest penetration rate in the world. Some of the differences are unexpected. In the United States, in March 2000, there were 185 Internet "host" computers per thousand people; in Japan, where consumers usually adopt new technologies almost as rapidly as do Americans, there were only twenty-three per thousand.[10] A gap has opened between the United States and Europe. Internet use is twice as high, per head of population, in America as in Europe; while the penetration of mobile telephones is 50 percent higher in Europe than in the United States.

Broadly, the take-up of new products will be influenced by three factors.

- **Culture**. If people are used to doing things one way, they may be slow to change – especially if, like one-fifth of the world's population, they are over sixty-five. Some developing countries, with their younger populations and fewer preconceptions, may thus be quicker than the older West, and especially Western Europe, to spot and take advantage of new possibilities.

- **Convenience**. Some technologies, such as the mobile telephone, save time. Others, such as the television, absorb it and are thus more vulnerable to competition from other time-consuming uses of communications technology, such as a long telephone call to a friend or a session playing on-line games. Buying on-line may not be more convenient than buying on the telephone. In the first case, the buyer does much of the work; in the second, the seller. Furthermore, the greater the effort it takes to master the use of a technology, the slower and more limited diffusion of that technology will be. Hence the PC will never be as widely used as the television or the mobile telephone.

- **Cost**. How fast a new technology is adopted will depend partly on how much and on what basis people pay for it. In the United States, where local telephone calls are largely unmetered, the Internet has flourished more than it has in Europe, where traditionally they have been timed. Conversely, the calling-party-pays system of charging for mobile telephones has boosted their use in Europe, while in America, where recipients still frequently pay for calls, the use of cellular phones has grown more slowly.

The way people (and organizations) trade off these three factors will vary from one country to another and decide the success of new technologies.

It takes time for any new technology to find its true markets. Demand for the home computer boomed once people began to use it as a superior games machine. Just as, with the arrival of electricity, people initially stood machines beside the power rather than developing the moving production line that transformed mass production, so today's

Internet users, individual and corporate, use this new medium largely for the same old services. Some of the technology's potential, such as its ability to allow machines to "talk" to one another or to use the world's PCs as a vast storage vault, has hardly begun to be explored. In spite of the rush of "minor improvements and subsidiary discoveries" that the dotcom stock boom helped to finance, completely new products and services are probably waiting to be discovered.

The Potential

The book's next three chapters look at the main areas of commercial and economic life that the death of distance will affect: consumers, commerce and the shape of the company, and the economy itself.

Consumers. The chief beneficiaries of this revolution are consumers. In businesses as different as music and travel, investment and gambling, the communications revolution is introducing new competition, holding down prices and multiplying choice. The greatest transformation will be seen in products that can be not just ordered but delivered on-line. Here is the biggest opportunity, not just for cutting costs and benefiting from global markets, but also for developing entirely new products and processes.

Consumers can also develop new relations with companies, if they want to. Companies can more easily track consumers' tastes and behavior and will be able to work harder to retain existing customers by studying their preferences. They can offer customers products tailored to their exact requirements. Indeed, they can involve customers in the development of products so that they become partners in the creative process of innovation.

Commerce. Consumers will be the main beneficiaries of the revolution – but companies will make more use of the Internet than individuals. Whatever their size, companies will have access to global markets. One effect of the increase in market size is, as Adam Smith, father of modern economics, observed two centuries ago, an increase in specialization. The computer's power to search, identify, and classify different tastes and needs will offer many more opportunities to niche businesses. Indeed, this will be a world of infinite niches, many of them occupied by very small firms earning a living from supplying an unusual product or service to a global market.

Many barriers to entry will fall, thanks to the declining cost of marketing and distributing "weightless" products on-line. But other barriers will rise. So, while there will be more corporate minnows, there will also be more whales. Brands will become more important, not less, as consumers struggle to pick their way through the rising torrents of information. In addition, many businesses that are based on a network – including hotels and banks – already find that communications amplify a network's value.

The shape of companies. On-line trading among companies, it is already apparent, will grow larger and alter more rapidly than retail commerce. Already, transactions between businesses account for about four-fifths of e-commerce revenues.[11] But because the first encounter most people have with the Internet is in the context of business-to-consumer commerce, they tend to underestimate its power to cut corporate costs and transform corporate structure.

The Internet brings a sharp fall in the costs of handling and transmitting information. Almost every business process involves information in some form: an instruction, a plan, an advertisement, a blueprint, a set of accounts. All this information can now be handled and shared more cheaply and swiftly. So many of the paper-shuffling tasks that employees now perform to keep a company going will either vanish or require fewer resources.

Electronic communication greatly enhances the ability of corporate managers to respond quickly. It slashes delivery times, whether for a car, a medical diagnosis, or a tax return. It also allows managers to monitor whether a job has been done on time and to specification. This may increase reliability and trust. That in turn represents an increase in efficiency, and the removal of a source of friction in complex modern societies. It also makes it easier for large companies to provide a truly local service tailored to specific needs and tastes.

Companies are already racing to redesign their supply chains – copying the pioneering examples of Dell Computer and Cisco Systems – and to build new electronic marketplaces in which to buy and sell. The result of intensified competition will be a greater emphasis on performance, productivity, and waste reduction. These benefits will occur, not only or even mainly in the dotcom companies launched in the heady late 1990s, but in most ordinary bricks-and-mortar businesses. As they spread, people will stop talking about "Internet companies" – indeed, the phrase already has an out-of-date ring. Every company will be an Internet company, just as every company is a telephone company. The Internet will become, in the phrase of Microsoft's Bill Gates, a company's "digital nervous system."

Outsourcing – of services as well as goods – will continue to grow. As communications improve, it will become ever easier for companies to track spare capacity and low prices. A knowledge-based company can buy in more of what it needs – design, say, or marketing or packaging – than can a traditional, "old economy" company. It can thus grow more integrated horizontally, rather than vertically. There will be new opportunities to bring together customers and suppliers, using the corporate communications network as the connective tissue.

Employees – suppliers of labor – may also develop different relationships with companies, working in loosely formed teams that contract for particular projects within a large company. Employees in different countries or regions will frequently cooperate on the same project. Some companies may become networks of independent workers, specializing in what they do best and buying in everything else, rather as a Hollywood studio employs relatively few staff, to provide banking and distribution services, and buys in most of the people who work on a movie. Many workers may not even be employees at all, just groups of freelancers selling their skills by the project or the month.

Manufacturers will increasingly think of themselves as service providers. They will discover ways to retain lasting links with their products. Car companies, for instance, will electronically track, monitor, and learn about their vehicles throughout an automobile's life cycle. New opportunities to provide services for customers will emerge, and some manufacturers may accept more responsibility for disposing of their products at the end of the cycle.

Politics and government will be transformed by free communications, changing the balance of power between governments and their citizens. People will become better informed, even if they live under dictatorial regimes. They will have new ways to make common cause. That will give new muscle to special-interest groups, but also empower grassroots activists in countries where democracy is fragile.

The restructuring of companies will have its parallel in a slower but even more important restructuring of government. Within firms, the old boundaries between departments will alter as companies turn to focus more intently on customers and their needs. The same reorientation will eventually take place in government, as the old alignments between departments are eroded. Where strong politicians force through the process, citizens will benefit enormously, for this revolution is essentially about increasing the efficiency with which services are delivered. Too often, bureaucrats will fight a long rearguard action to preserve their traditional "silos" intact.

The authority of the nation state will also be affected. The death of distance will erode national borders, curbing sovereignty. Governments will be forced to cooperate more than ever before – with one another, and with the private sector. Tax bases will erode. As the borders of large countries grow more porous, the handicaps that affect small countries will diminish. In a world of infinite electronic niches, the niche nation will have space to flourish.

Economies. The vast diffusion of knowledge and information, the basic building blocks of growth, will undoubtedly bring economic benefits. Economists vigorously debate the size, timing, and extent of those gains. But one effect will be to cut costs; another, to allow the capacity of the global economy to be used more fully and efficiently. In addition, this revolution clearly fosters innovation – the introduction of new production methods, new products, and new kinds of industrial organization – which is the main force driving growth and thus living standards. The communications revolution speeds up the diffusion of innovation. It will allow companies to react quickly and individuals to spot opportunities.

It will also exert downward pressure on prices – and on profits. Information is essential to make markets work well. Many more companies will have access to accurate price information, enhancing competition and helping to curb inflation. But it will be harder than ever to make comfortable profits for any length of time. That is good news for consumers, but makes the dotcom market fever of 1999 look foolish. True, there will be profitable niches to exploit. For many companies, however, the "new economy" means, not supernormal profits, but "profitless prosperity."

The problems

Against such potential should be set the problems and policy issues that will face this altered world. Two chapters address these questions, many of which revolve around the powers and responsibilities of government.

One set of issues is whether innovation justifies a certain amount of monopoly, a point raised most strikingly by the Microsoft case. In October 1998, America's Department of Justice launched an antitrust case against the world's largest software company which resulted, in April 2000, in a ruling that the company was an "abusive monopolist" that should be broken up. Some of the industries of the "new economy" have characteristics that make them particularly prone to concentration, such as their reliance on networks and standards. Many others want government to protect them from competition through patents, copyright, or other legalized kinds of monopoly power. The reason is that they have high development and launch costs but extremely low costs for producing each additional item for sale. To recoup their development spending, they demand much more for their products than each costs to produce – as everyone who has ever growled at the price of a CD or a package of software can testify.

A second set of issues concerns what governments should do, and can do, to counter the increased accessibility of evil that comes along with more useful knowledge. The globalization of communications will make it harder to enforce national laws to protect children, preserve privacy, or prevent terrorism. Part of the price of freedom will be a greater need for individuals to take responsibility for their own lives and those of their families, and for the smooth running of society.

Finally, **society** will be affected, with the disappearance of the old demarcation between work and home. The office may become the place not for routine work, which will take place at home or on the road, but for the social aspects of a job such as networking, lunch, and catching up on office gossip. The result will be to invert the familiar roles of home and office. Social life itself will become more flexible. For young people, the combination of the mobile telephone, global-positioning satellites, and a sort of instant messaging will enable them to locate friends, switch plans, and evolve ever more complicated personal arrangements.

"The only thing I'd rather own than Windows is English or Chinese or Spanish, because then I could charge you a $249 right to speak English and I could charge you an upgrade fee for when I add new letters like 'n' or 't'," says Scott McNealy, boss of Sun Microsystems.[12] English is almost as essential to the smooth running of the Internet as Windows is to the PC. Fortunately it is public property, for it has become the world's default language, the standard medium of Internet commerce. Does this spell doom for other languages? On the contrary: the infinite shelf space on the Internet provides room for any number of languages and for their speakers and writers to communicate. New communities of interest and culture will emerge, linking the world's scattered people horizontally, as it were, by taste, rather than geographically by neighborhood.

Fear and doubt

For many people, this prospective new world is distasteful or frightening. They worry about society's vulnerability to technological breakdown and computer crime. The more society relies on technology, many argue, the bigger the problem when something goes wrong and computer systems are hacked into or go haywire. Pessimists see the prospect of many jobs destroyed and a few sets of skills disproportionately rewarded. Others squirm at the crassness of the mass culture that low-cost electronic distribution conveys, at the erosion of old values, and at the apparent "Disneyfication" of so much that once distinguished one nation from another.

Others again dislike the deluge of electronically delivered information. They resent the threats to their privacy. They chafe at the demands on their time. With some reason: the volume of messages addressed individually to each person in the United States has increased about a thousand-fold in the past two centuries.[13] The doubters fear that the personalization of communications will go hand-in-hand with social fragmentation, and that the institutions that have held society together in the past, whether universities or local stores, will vanish, to be replaced by "learning-in-a-box" or "store-on-a-screen."

In fact, many of these worries already apply in the non-electronic world. The water supply, the freeways of Los Angeles, the Tokyo subway

– all make society more vulnerable than before to sudden disruption, but in exchange for benefits so vast that we accept the risks. And against the danger of jobs destroyed should be set new opportunities, including chances for some who lacked them in the pre-electronic world. "On the Internet," said a famous New Yorker cartoon, "nobody knows you're a dog." Nobody knows either whether you are young or old, black or disabled, or a woman.

The threats to privacy and time are more disturbing. Privacy is gone in an electronic era, but then privacy is a novel concept, a construct of large cities and the rich world. In the villages and towns that our ancestors lived in, privacy was rare. The difference is that then, it was neighbors who knew your affairs; now, it will be mainly strangers.

As for the assault on our time and attention, the real divide now between haves and have-nots is between those with too much work and those with too much leisure. As people spend more and more of their lives in retirement they may welcome ways to fill their time. Those who lack spare hours and minutes, the scarcest of all commodities, will look for technologies and companies that save time, by filtering information or by making it possible to use the "downtime" of commuting or traveling.[14]

Hope and opportunity

Who will be the winners from this revolution? Developing countries, for a start. Already today's emerging economies account for a quarter of the world's telephone traffic, according to the International Telecommunication Union (ITU), an official club of the world's big telephone companies. By 2005, that will rise to half. By then, China will be one of the world's top three countries for international calls. India will be close behind. In addition, the countries of Asia and Latin America where telephone penetration is already relatively high will almost certainly see the fastest adoption of the Internet.

Many developing countries will use new technology to race into the new century. They will build wireless telephone networks (fixed as well as mobile) rather than wired ones, and they will have access to the Internet through mobile telephones long before they acquire PCs. They will find the potential gains are vast. The new economy needs less basic infrastructure than the old one. It takes a decade to repair the road from Nairobi to Mombasa and it would take even longer to clear the Nigerian waiting list for landlines. But wireless offers a quick, inexpensive alternative to traditional, wired telephone networks. Not surprisingly, Africa in 1999 had the fastest growth rate in mobile telephone subscribers in the world – 103 percent – and some African countries, including Uganda and Côte d'Ivoire, now have more mobile than fixed connections.[15] Problems with the fixed-line telephone network have already made e-mail, with its useful time-shifting ability, the default method of communication in many parts of Africa.

As they adopt new communications, developing countries have more potential than rich ones to increase productivity. Give an Indian truck driver a mobile telephone, or a Chinese village an Internet connection, and you create a way for them to leapfrog decades of development. You open the door to information from the rest of the world that was previously accessible only to a few: technological information, which can enable developing countries to catch up with the rest of the world in industry, health care, education, and a host of other services; commercial information, raising aspirations and transforming patterns of consumption; and political information, that will foster the spread of democracy and open discussion of public affairs. As investments go, spending on communications in the developing world is formidably productive.

However, inexpensive communications alone will not be enough to narrow the gap between the poor world and the rich. To allow communications to work their magic, poor countries will need sound regulation, open markets, and, above all, widely available education. Where these are available, countries with good communications will be indistinguishable. They will all have access to services of first-world quality. They will be able to join a world club of traders, electronically linked, and to operate as though geography had no meaning. This equality of access will be one of the great prizes of the death of distance.

Change is always unsettling, and we are living through one of the fastest periods of technological change the world has known. But at the heart of the communications revolution lies something that will, in the main, benefit humanity: global diffusion of knowledge.

This is a revolution about opportunity and increasing human contact. It will be easier than ever before for people with initiative and ideas to turn them into business ventures. It will be easier to learn new things, acquire new skills. Above all, it will be easier to find someone to talk to – to communicate, whether with friends or strangers, relatives or customers. As a result, the world will, probably, be a better place.

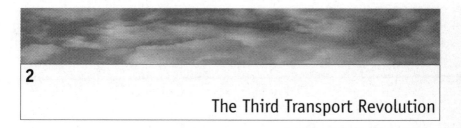

The Third Transport Revolution

In 1798, Thomas Malthus, an English cleric turned economist, wrote a gloomy and influential essay. "Population, when unchecked," he observed, "increases in a geometrical ratio. Subsistence increases only in an arithmetical ratio...This implies a strong and constantly operating check on population...No fancied equality, no agrarian regulations in their utmost extent, could remove the pressure of it even for a single century." Like so many economists after him, Malthus was grandly and utterly wrong. The population growth of Europe in the century that followed was not starved into stability. Instead, it rose faster than at any time before or since.

What wrong-footed Malthus was not just a rise in Europe's farm productivity. More important was the first great transport revolution. The coming of steam, and its use in ships and railways, made it not just technologically possible but economically attractive for the New World to feed the old one. Ocean freight rates fell faster than prices for much of the nineteenth century, and long-haul rates fell more than short-haul ones. The effect was to reduce the price of food and other commodities and to encourage the development of new lands. As Douglass North, a distinguished economic historian of the period, puts it, "The declining cost of ocean transportation was a process of widening the resource base of the Western world."[1]

In the twentieth century, the great changes in transport technology grew from the refinement of the internal combustion engine and the jet engine. The car did something that really successful new technologies often deliver: it brought to the many something that had previously been reserved for the few. Other inventions of the late nineteenth century, many apparently modest, did the same. The paper dress pattern allowed ordinary women to make themselves passable copies of high fashion; the lawnmower enabled the suburban masses to have miniature copies of the sweeping swards of stately homes. The car provided even more personal mobility than was formerly available to those who could afford the palaver of keeping a horse.

Both revolutions hold lessons for the third great transport revolution, which is transforming the cost of communicating ideas and knowledge. They show how frequently, since Malthus, people have misjudged the technological future and thus its consequences. They show how broad the effects of technological change can be once it is incorporated into an innovation that diffuses across the whole economy, as electricity did a century ago. And they demonstrate the important connection between technological change, convenience, and pricing. Because even quite inept people can learn to drive a car (although it is the most complex mechanical task that most folk undertake), they can substitute their own labor for the bought-in labor that drives a bus or a train. The popularity of the car is the result of the powerful combination of convenience, flexibility, and cost that has risen more slowly than that of public transport.

Changes in transport technology, driving down costs, are among the most potent forces shaping human lives. [Fig 2-1] They have opened continents, transformed living standards, spread diseases, fashions, and folk around the world. The work from which this book draws its

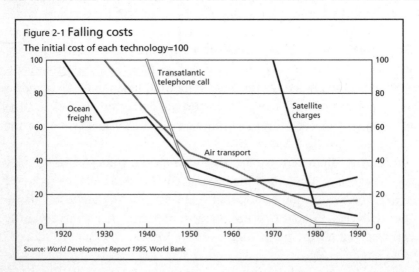

Figure 2-1 **Falling costs**

The initial cost of each technology=100

Source: *World Development Report 1995*, World Bank

title, Geoffrey Blainey's famous study of the way distance shaped the history of Australia, shows the effect of the twelve thousand miles that separate the country from Western Europe, the source of most of its people and ideas. Long-distance transport is the essence of Australia. It "illuminates the reasons why Australia was for so long such a masculine society, why it became a more equalitarian society than North America, and why it was a relatively peaceful society."

The railway in particular changed the way people thought about their world. Before the railways came, many places had their own time zones: in Britain, by the 1850s, the pressure of the railway timetable made London time standard throughout the country. Trains brought with them a new need, not only for a single time but also for punctuality. They altered patterns of employment and trade, cutting distribution times and costs. And railway shares, first booming and then slumping, gave the new middle class of the Victorian age a taste of the excitement and peril of investing in communications technologies that their great-grandchildren experienced many years later.

Great though the impact of change in physical transport has been, changes in the transport of ideas and information have done more. By spreading knowledge, they widen the base of economic growth.

Early Days
· · · · ·

Each new development in the technology of communications in the past two centuries has made communication faster and cheaper. Each offers lessons for the revolution now under way.

Take one of the most effective pieces of government intervention in communications, the 1840 reform of the British postal system by Rowland Hill. It offers powerful lessons for the way the pricing of modern communications is structured. Before Hill's reforms, the distance a letter travelled determined the price of the stamp, and the payment was made by the recipient, not the sender. This led to nasty surprises: Hill first thought up the reform when he saw his mother unable to pay the fee. His Penny Post set a single, universal rate for letters sent anywhere in Britain, paid by the sender.

Rowland's reforms showed how sensitive communications are to their charging system. Within two years of the coming of the Penny Post, the volume of letters doubled.[2] The success of a single, simple tariff for the whole of Britain was, in a sense, an early application of the principle of the death of distance to communications pricing. It was also, in effect, a "calling-party-pays" system of charging for the mail. There are important implications here for the United States, where a reluctance to pay for incoming calls on mobile telephones encourages many people to keep them switched off and to use them only for outgoing calls. In most other countries, the cost of mobile telephone calls is paid by the caller.

The price of letters in the early nineteenth century holds further lessons for communications, observed by Andrew Odlyzko, head of the mathematics and cryptography research department at AT&T Laboratories.[3] In the United States, revenue from the mail service subsidized "to an extraordinary degree" the distribution of newspapers by post – because the government, eager to inform the citizenry, decreed that it should be so. This suggests two things. First, government's desire to use cross-subsidies to widen public access to communications technologies may not reflect what people would choose to do with their own money. Second, people are willing to spend more on point-to-point communications than on entertainment or information – both of which are commodities many people expect will finance the development of the Internet.

The evolution of the Victorian post office also shows how pervasive the effects of changes in communications can be. One consequence of the Penny Post was the insertion of letterboxes into front doors, because postmen no longer needed to knock to be paid. Another was the introduction of pillar boxes, because letters no longer had to be taken to a central receiving office. Both reforms are credited to Anthony Trollope, the Victorian novelist who worked under Rowland Hill. In her biography of Trollope, Victoria Glendinning points out that the pillar box struck a blow for the liberation of Victorian women because it "gave freedom to over-protected girls to carry on private correspondences," a freedom of which Trollope's young heroines took full advantage.[4]

Economic life changed too. Merchants in America's Eastern cities used the nineteenth-century mail to gather information, enraging far-off cotton growers and farmers who found that New Yorkers knew more about crop prices than they did. In the American debate over slavery, the mail offered abolitionists a low-cost way to spread their views, just as the fax machine and the Internet have cut the cost and widened the scope of political protest and lobbying. The post helped to integrate the American nation, tying the newly opened West to the settled East.

The post has also shown the power of an established communications technology to survive and adapt to the coming of rivals (for the one important exception, see page 26). Not only is the volume of mail carried by most of the world's postal services still rising; the coming of the Internet has bolstered the growth of courier services, which are still the fastest way to take physical objects from one place to another. Ideas may move at the speed of light. Goods never will.

The Victorian Internet[5]

Together with the code that the American Samuel Morse had perfected by 1838, the electric telegraph brought the greatest change in the speed and cost of communications in all of history. It marked a far more dramatic wrench with the past than did the reorganization of the postal service. Indeed, it was a truly disruptive technology.[6] Before its commercial introduction, along a stretch of railway out of London in 1839, the thirst for speedy communications had been best satisfied by the mechanical telegraph, a tower with movable arms whose positions could be read from afar with a telescope,[7] or by courier services such as the Pony Express, or by homing pigeons. (Charles-Louis Havas, whose name is perpetuated by a large French education and software publisher, was one of the proprietors of some twenty-five thousand pigeons kept in Antwerp in the 1840s to fly news of shipping and trading to the markets of Europe.)[8]

The electric telegraph burst upon the world as suddenly as the Internet has done, and was greeted with a similar mix of hype and gloom. In 1858 it took forty days for news of the Indian Mutiny to reach London; by 1870, several telegraph lines connected India to London and news of a problem with the tea harvest would affect London markets within hours.[9] Victorians spoke in awed tones of "the annihilation of space and time." Unlike anything that had gone before, the telegraph changed the world and the way people thought about it.

As with the Internet, the telegraph was built at astonishing speed and with massive private investment. At the beginning of 1861, there were 1,120 kilometres of submarine cable; ten years later, 46,000 kilometres; twenty years later, 147,500 kilometres. The cumulative cost of the first successful transatlantic cable was $12 million – almost double the $7.2 million cost of buying Alaska from Russia in 1867.[10]

Many other aspects of the telegraph carry hints of the Internet to come. Thus the telegraph rapidly transformed commercial, social, and political life – sometimes in strikingly similar ways. It produced the same string of complaints about information overload. W.E. Dodge, a New York businessman, complained in a speech in 1868 of the harm that daily telegramed market reports did to the businessman's digestion and domestic life. "The merchant goes home after a day of hard work and excitement to a late dinner, trying amid the family circle to forget business, when he is interrupted by a telegram from London, directing, perhaps, the purchase in San Francisco of twenty thousand barrels of flour, and the poor man must dispatch his dinner as hurriedly as possible in order to send off his message to California."[11] It all sounds dismally familiar.

For other businessmen, though, the telegraph was an immense convenience. Alfred Marshall, a British economist whose *Principles of Economics* is still one of the best accounts of the role of innovation and technological change in economies, describes how the combination of the telegraph with the railway and the manufacture of interchangeable parts for machinery enabled farmers in distant places to work more efficiently than ever before. "A farmer in the North West of America, perhaps a hundred miles away from any good mechanic's shop, can yet use complicated machinery with confidence; since he knows that by telegraphing the number of the machine and the number of any part of it which he has broken, he will get by the next train a new piece which he can himself fit into its place."[12]

Like the Internet, too, the telegraph enormously improved inventory control, especially in distant markets. Geoffrey Blainey records how Australia – the most distant market of all – was periodically flooded with gluts of boots or mirrors in the days before the telegraph, because British exporters had to guess the needs of a market on the other side of the world. Ignorance at the Australian end imposed even greater penalties. South Australian copper mines "could be vigorously producing copper because the most recent news from London indicated that the world price of copper was high, when in fact the price of copper in London was so unfavorable that their mine was working at a loss."[13] For the village producers of the developing world, the telephone has only recently brought the same vital ability to follow national or global market prices for their key commodities.

"The telegraph should be an instrument of politics, not of commerce," snorted France's minister of the interior in 1847. His irritation is echoed by French distaste for the Internet: in both cases, France got there first, but with the wrong technology. In pioneering the mechanical telegraph in the 1790s, and in developing Minitel, an electronic information service owned by France Télécom and launched in 1983, France beat the rest of the world but then headed off down the wrong track.

In fact, the telegraph did become an instrument of politics. It gave the British government a tool with which to administer its sprawling colonial empire. It also gave Britain a handy weapon against the French. During the second half of the nineteenth century, Britain dominated the ownership of the cable industry and the business of manufacturing submarine cables, just as the United States now dominates the Internet and the manufacture of fiber-optic cable. Because many cables were routed through London, the British government routinely intercepted telegrams about French colonial matters, to the understandable indignation of the French.[14]

The telegraph had its hackers and its swindlers, its romances and its tragedies, some of them eerie precursors of later electronic dramas. In 1874, after aborigines attacked a telegraph office in Australia's Barrow Creek, a man lay dying. The news of the attack was telegraphed south to Adelaide, where the wife of the dying man was hurried to the office. There, according to a journalist of the day, she heard "the exhortations by wire of her husband – distant twelve hundred miles, the wire at his very bedside – each bidding an eternal adieu to the other by the click of the instrument."[15] In the course of 1996, two stranded climbers on Mount Everest used satellite telephones to call their wives. One wife, two thousand miles away in Hong Kong, was able to arrange her husband's rescue; the other, sadly, could merely say a last farewell.[16]

Unlike any other important communications technology, the telegraph died out. It flourished well into the twentieth century – indeed, 1945 was the peak year for the sending of domestic telegrams in the United States.[17] But when the war ended its use dwindled. What killed it? Two things. First, it remained a relatively expensive way to send a message. In the 1870s, a twenty-word message still cost the equivalent of around $200 in today's prices to send.[18] More important, it ran up against a technology that did everything the telegram could do but allowed genuine interactivity: the telephone. The telegram had to be physically delivered to the recipient by a lad on a bicycle; the telephone, like the car after it, allowed an individual's own labor to replace bought labor.

Wireless

Sending telegrams without wires to carry them came early in the twentieth century. Legend has it that the facility was a hit with the wealthy passengers of the *Titanic*, who received so many calls from America that the wireless operator broke off conversations with the nearby California that might otherwise have saved the passengers on the doomed ship. Transatlantic telephone calls also went by radio until the first submarine cable to carry them was laid in 1956. In the early twentieth century, AT&T saw radio as a potential threat to its landline network. Not until the final years of the century did that threat become real, as wireless became a way to carry billions of telephone calls and messages from point to point, rather than merely a means to broadcast mass entertainment.

Radio's explosive growth as the world's first big electronic entertainment medium tells four stories that still resonate today. First, the fastest take-up was in the United States. While most countries, with state-owned telegraph and telephone services, moved on naturally to state broadcasting, the only state-owned communications service in America has been the mail. America's ferment of competition in radio in the 1920s sparked a boom in radio shares quite as mad as the Internet bubble of the late 1990s. The young private networks experimented with different kinds of finance. At first, their broadcasts were subsidized from sales of radios: give the punters good content, the argument ran, and they have a reason to buy the set. The radio commercial came later.

A second echo can be found in the power of communications to change commerce. National brands, which national magazines had begun to build, flourished on the back of national broadcast entertainment. As commercial radio, and then television, arose, so did whole new industries: advertising agencies, program makers, public-relations consultants, stars. Hollywood's bread and butter has long come from television; films are merely the jam.

Third, radio offers a fascinating example of the way one technology (and especially one transport technology) can determine the fate of another. In many countries, more people now listen to the wireless in their cars than in their homes. Time at the wheel may turn out to be the most important slice of the packed human day that communications companies compete for.

Fourth, huge though the revenue from radio and television might be, it is modest compared with that from the telephone, the unbeatable tool for personal communications. Indeed, the revenue generated in the United States from the cellular telephone alone overtook, in 1998, the entire revenue from television and radio. The big money, it would seem, is in point-to-point communications, not broadcast.

The Roots of Revolution

All three of today's fast-changing communications technologies have long histories. The telephone was invented in 1876, the first television transmission occurred in 1926, and the electronic computer goes back to 1946, if not earlier. For much of that time change has been slow, but in each case it began to gather pace in the late 1980s.

The telephone

One of the curiosities of the communications revolution of the twenty-first century is that it has, at its heart, a technology that goes back to the nineteenth. The telephone was greeted initially with almost as much skepticism as the telegraph had been. "Too many shortcomings to be seriously considered as a means of communication. Inherently of no value to us," said Western Union in a dismissive memo after being offered Alexander Graham Bell's patent for $100,000. Even in 1905 the Bell System in America still estimated that a telephone for every five Americans would be saturation point.[19]

The mistake was to recur when the mobile telephone arrived. Cellular communications date back to the period immediately following the second world war, but became commercially viable only in the early 1980s when the collapse in the cost of computing made it possible to provide the necessary processing power at a low enough cost. But when AT&T asked McKinsey, a consulting firm, how many mobile telephones would be in use in the world at the turn of the century, the consultants thought that in the United States there would be fewer than a million. AT&T pulled out. Today, a million or so new subscribers around the world sign up for a mobile telephone every three days.

It is almost as easy to head unwittingly into a dead end as to fail to take the right turning. As with the Internet today, it was not at first clear what the best use for the telephone would be. A service called Telefon Hirmondo was set up in Budapest in 1893 to provide what one might call on-line news and entertainment, interspersed with advertising. In the evening it offered a children's program and lessons in English and French, on Sunday a grand concert. At its peak, in the 1920s and 1930s, the service had more than ten thousand subscribers, and it kept going until 1944.[20] But mass communication by electrical means turned out to be the wrong use for telephone lines.

Just as with the Internet, the speed of take-up of the telephone varied from one country to another – and in a remarkably similar pattern. No country grabbed the telephone in its early years faster than Sweden, which in the 1890s had more subscribers per head of population than anywhere else. Close behind came Denmark: its telephone use had caught up with Sweden's by the start of the first world war. Germany and Britain trailed behind, but were still well ahead of France and Italy. That is much the order for the uptake of cellular telephones and the Internet in Europe today: Sweden has one of the highest rates of adoption in the world – with more than half the population describing themselves as Internet users[21] and more than 60 percent owning a mobile telephone[22] – while France and Italy trail behind Britain and

Germany. Clearly the willingness to adopt new communications technologies is not just a matter of wealth or language or even state involvement. Perhaps it has more to do with climate, culture, and the length of
winter nights. [Fig 2-2]

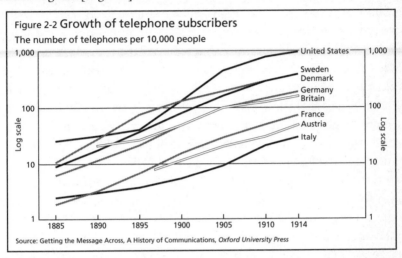

Figure 2-2 **Growth of telephone subscribers**
The number of telephones per 10,000 people

Source: Getting the Message Across, A History of Communications, *Oxford University Press*

Throughout the twentieth century, though, it was the United States
that topped the league for telephone use. In 1997 Americans used their
telephones for just under thirty-seven minutes a day. By then, of
course, the telephone's use had begun to reflect the growth of dial-up
Internet access. But even in Sweden, where Internet penetration is
greater than it is in the United States, the average person made only
just under twenty-one minutes of calls a day. In Britain, the figure was
less than thirteen minutes – just over one-third the use recorded in
America. One reason for America's century-long predominance is
almost certainly the level and structure of telephone charges. Not only
have charges generally been lower overall in America than elsewhere
(in 1997, the industry's revenues were scarcely higher, as a fraction of
GDP, than they were in Britain from roughly one-third as many calls),
but, in addition, America's system of "free" local calls (a flat monthly
fee for subscribing to the telephone network but none, in most places,
for placing a local call) was in place right from the start of the twentieth century. It almost certainly encouraged Americans to use their telephones more than people in other affluent countries do.

Most people – not just Americans – have historically used their telephones mainly to call folk who live nearby. One reason is that the tele-

phone network had, throughout most of its long existence, the least capacity for what should have been its most irreplaceable service: long-distance communication. For a century, there was little capacity even on the most important of routes, across the Atlantic. On the eve of the second world war, there were only four (wireless) commercial circuits between the United States and Europe, three to London and one to Paris. Even after a submarine cable replaced the radio link in 1956, it could carry only 35 telephone conversations. So scarcity continued. Walter Wriston, former chairman of Citibank, recalls what it was like to be an international banker in the 1950s and 1960s. "It could take a day or more to get a circuit. Once a connection was made, people in the branch would stay on the phone reading books and newspapers all day just to keep the line open until it was needed."[23]

It was not until the late 1980s that capacity on the principal long-distance routes began to explode, as it became possible to send a signal along a fiber-optic cable made of glass so pure that a sheet seventy miles thick would be as clear as a windowpane. The first transatlantic fiber-optic cable, with capacity to carry nearly forty thousand conversations, went on-line only in 1988. The cables being laid in the twenty-first century can carry some three million conversations on a solitary strand of fiber, the width of a human hair.

Only as capacity grew did the premium for long-distance calls begin to fall rapidly – generating the demand that encouraged more investment. Nowadays, capacity is so plentiful on some routes that domestic telephone calls made during the morning in Britain are switched through the United States, when Americans are asleep but Britons are making most use of the telephone. The cost of carrying an extra telephone call across the Atlantic, and on many other long-distance routes, is virtually zero.

By 2000, vast investment in "backbone" networks, travelling across continents and under oceans, had created a massive glut in fiber-optic capacity. But that glut may turn out to be short-lived. Increasingly, the fibers are filled, not with the chatter of human voices but with the bits and bytes of data traffic generated by the Internet, which doubles in volume every year. Telephone talk makes fairly constant and predictable demands on capacity; the Internet, by contrast, is "bursty," in the vivid phrase of engineers, making sporadic demands of great intensity. But capacity will remain ample enough for the cost of long-distance communications to continue to fall, changing our mental map of the world and providing the drive behind the death of distance.

Television

Few inventions in history have been adopted as fast and as pervasively as television. From the launch of America's first commercial television station in 1947 to the moment when half of all the country's homes had a set took a brief seven years.[24] Globally, a mere eight thousand homes had television sets at the end of the second world war. By 1998, that number had risen to 938 million – almost three-quarters of the world's households. Not only do far more people thus see television than have access to a telephone; television also absorbs more time and wields more influence. Most of the hard-won gains in leisure time in the second half of the twentieth century were matched by rather more easily won increases in the time people spent watching the box.

As with the telephone, the basic technology of the set hardly changed for most of television's history. The telephone offered few new products until mobility came along. The main exception was the fax machine, a way to send data across a telephone line (as the Internet does now in far greater quantities) and a fine example of the democratizing power of successful innovation. It provides individuals and small businesses with a service that was once too expensive and too cumbersome for any but the largest firms.

In the case of television, the most remarkable change was the development of communications satellites, which probably did as much to transform people's view of the world in the twentieth century as the railways and telegraph had done in the nineteenth. In 1963, thanks to the launch the previous year of Telstar, the first private communications satellite, people everywhere witnessed for the first time an important but distant political event as it was taking place: the funeral of President John F. Kennedy. The psychological impact was huge. This unprecedented new link between countries created a sense that the world's peoples belonged to a global, not merely local or national, community. Over the subsequent decades, satellite transmission transformed television. As recently as the 1970s, more than half of all television news was at least a day old. Today, almost all news is broadcast on the day it occurs.[25] Big events – the fall of the Berlin Wall, the funeral of Diana, Princess of Wales, the war in Kosovo – go out to billions of viewers as they happen.

Two other changes began, in the final few years of the twentieth century, to transform what viewers receive. First came choice. Until the 1990s, most television viewers around the world had access to no more than half a dozen television channels – and often to only two or three. The main reason was purely physical. Most television in most countries is broadcast over the air, and analogue signals are greedy users of spectrum. Only in the United States and a handful of other countries have cable-television networks – less constrained by the limits of spectrum – brought people real viewing choice. When, at the end of the 1980s, communications satellites began to broadcast directly to a small dish attached to people's homes, television's infrastructure costs fell sharply, multichannel television became widely available, and choice expanded with breathtaking speed.

In the late 1990s came another potentially revolutionary change: broadcasters began to transmit television in digital, not analogue, form, allowing the signal to be compressed. As a result, far more channels can be transmitted, whether from satellite, through cable, or even over the air. Like the long-distance parts of the telephone network, a service that had been constrained by capacity shortage for most of its existence suddenly began to build more capacity than it knew what to do with.

The networked computer

The newest of the three building blocks of the communications revolution, the electronic computer, has evolved the fastest. In 1943 Thomas Watson, founder of IBM, was making the familiar mistake: the world market, he forecast, had room for about five computers. As recently as 1967, a state-of-the-art IBM, costing $167,500, had a memory that could hold a mere thirteen pages of text.[26] In 1977 came another misjudgement, from Ken Olsen, chairman and founder of Digital Equipment Corporation. "There is no reason", he announced, "for any individual to have a computer in their home."[27]

Two astonishing changes have occurred since then. First, computing power has grown hugely. Second, computers are increasingly interconnected, principally through the Internet, in a network of vast potential power.

The increase in computing power has followed a principle known as Moore's Law, after Gordon Moore, co-founder of Intel, now the world's leading chip maker. In 1965, Moore forecast that computing power would double every eighteen months to two years, and so it has done for three decades, ever since Intel developed the first successful single-chip microprocessor – the brain and memory of a computer – early in the 1970s. Engineers have found ways to squeeze ever more integrated circuits of transistors on to small wafers of silicon (or "chips"). And Moore's law continues to apply. By 2010, computers will have ten million times the processing power that they had in the mid-1970s. As the power of the chip multiplies, so the price of computing power falls, computer size decreases, and computing capacity rises.

As a result, the computer has become a consumer durable.

An important moment came in 1977 when Steven Jobs and Stephen Wozniak, two young computer enthusiasts, launched the Apple II, opening the way for the computer to become a household good. Today, more than a third of households in the rich countries that belong to the Organisation for Economic Co-operation and Development (OECD) have a PC.

In addition, computing power is now embedded in all sorts of products – sometimes with unexpected results.

Sony's PlayStation2, coupled with a video camera, could make an ideal missile-guidance system. Japan's government slapped export controls on the device. The episode provided a vivid instance of the computing power installed in household gadgets: the PlayStation2's central processor has twice the power of Intel's most advanced Pentium chip.[28]

As for the PC, it took a couple of new technologies to turn it from a high-powered typewriter and games machine into something more: the Internet and the World Wide Web. Both demonstrate an important historical principle of new inventions – they never come from the industry they supersede. The railway industry did not invent the automobile. The telephone industry did not invent the Internet. Instead, like Western Union with Alexander Graham Bell's patent, it turned the novelty away.

In fact, the Internet is a product not of entrepreneurialism but of academic research, publicly financed. It grew from an experiment at BBN, a small company in Cambridge, Massachusetts, backed by the American Defense Department's Advanced Research Projects Agency (ARPA), to connect computers across the country as a way to share and so multiply their computing power. The experiment, envisaged as a response to the expense and scarcity of existing computers, created a nationwide network (ARPANET[29]) that initially linked university computers. Throughout the early 1980s, the main users were the large research universities. They strung together their networks of computers using Ethernet, a local area network that had been designed by Robert Metcalfe back in the early 1970s, and then hooked them together to form a network of networks that grew into the Internet. It was, beyond a doubt, the American military's most influential peacetime project.

Some of the Internet's most important characteristics spring from its early history.

A single standard. Because it aimed to link computers, each with different operating standards, the Internet has at its heart a way to overcome these potential incompatibilities. To talk to one another, computers need not just a physical link but also a common language. In 1974 TCP/IP (Transmission Control Protocol/Internet Protocol), which lays down the format in which all the data sent over the Internet is packaged, was designed (mainly by Vinton Cerf and Robert Kahn). It was introduced formally in 1983, the date usually taken as the Internet's proper starting point.

TCP/IP is the essence of the Internet: a common language and a set of rules that enables computers all over the world to talk to one another, whether they are PCs or Apple Macs, vast university mainframes or domestic laptops. Through the Internet, any number of networks – telephone networks, cable-television networks, or in-house corporate "intranets" – can connect and thus behave as a single network of networks.

Distributed and packet-switched. During the tense years of the cold war between the Soviet Union and the West in the early 1960s, Paul Baran, a researcher at the RAND Corporation, was interested in designing a communications system that could survive nuclear attack. Vulnerable communications were a threat to peace, he argued, because if communications could be easily disabled before retaliation to an attack could be launched, pre-emptive strikes became more likely. Baran came up with the idea of a distributed network designed like a fish net, rather than the centralized network then typical of the telephone system. If one link in such a network were to be knocked out, a message could travel by another route.

To make that easier, Baran came up with a second idea: split the message into fragments and send each one separately. Computers, unlike telephones and television, had always been digital, handling information in a stream of digital ones and zeroes. Baran envisaged a network of unmanned switches, or computers, reading the address on each message and sending it on – "hot-potato routing," as he called it – to the next junction on the network. At each junction the process would be repeated until the message reached its destination. There, other information encoded in the message would enable it to be reassembled into its original form.

AT&T – predictably, given companies' history of spotting the innovations that count – was slow to take an interest in Baran's ideas. But "packet-switching," as his innovation came to be called, has become the main way to send data around the world and the core of the Internet, the largest of all data networks.

A common good. The Internet's standards, such as TCP/IP and HTML (hypertext mark-up language), the standard coding system of the World Wide Web, are public property, non-proprietary, available for anybody to use freely. Because the Internet grew out of the worlds of public funding and academic research, its protocol can be used without license, payment, or permission. In that sense, it resembles the English language more than it does, say, Microsoft's near-ubiquitous Windows operating system, whose private owner makes money (lots of it) from its popularity. This wonderful open quality has stimulated a great surge of creativity and continues to pervade the Internet, as is shown by the rise of Linux, an operating system developed and refined by volunteers and available freely for anybody to download. While the telephone network is controlled by the telephone companies, which set rules about what can be connected to it and on what terms, the Internet puts power in the hands of the user.

The World Wide Web

Although the use of the Internet grew rapidly in the 1980s and early 1990s, doubling every year, it owes its transformation into a popular success to an invention of 1989. In that year Tim Berners-Lee, a British researcher at CERN's European Laboratory for Particle Physics in Switzerland, came up with a way of keeping tabs on CERN's vast resources of research and knowledge. The structure of CERN, he said, was a "multiply connected 'web' whose interconnections evolve over time."[30] The solution, he suggested, was hypertext: a way to cross-reference by linking a word or phrase in one document on a screen to related information that may be stored on a different computer or even in a different network. It should not be centralized and users should be able constantly to add links of their own. A year later, Berners-Lee created a program called a "browser" that allowed a user to look at information stored on networked computers called "servers." Early in 1991 the World Wide Web was born.

But to become popular with people other than research scientists, one more step was needed. In 1993 Marc Andreessen, a twenty-three year-old programmer at the University of Illinois, and his colleagues came up with Mosaic, the multimedia Web browser. It enabled Web pages to include images side by side with text for the first time, making them more attractive to look at and use. Berners-Lee was shocked. "This was supposed to be a serious medium – this is serious information," he reportedly complained.[31] But within two years the volume of Internet traffic involving Web pages went from almost nothing to more than half of the total.

"Berners-Lee was like Thomas Edison, who thought that the phonograph he'd invented was for dictating office memos," argues John Naughton, a recent historian of the Internet. "Andreessen and co were closer in spirit to Emile Berliner, the guy who realized that the killer app. for the phonograph was playing pre-recorded popular music."

The Web is now often spoken of as synonymous with the Internet itself – understandably, for the Web made the Internet easily accessible and fun. It offers colourful pictures, music, and moving images as well as data and text. Point a mouse at a hypertext word or picture and click, and a new file magically appears, perhaps stored on a computer on the far side of the planet. A toddler can do it.

Grown-ups, however, needed one more thing: the power to search. In 1994 two Stanford University electrical engineering graduates, David Filo and Jerry Yang, created an on-line directory for Web sites and called it Yahoo! (standing for "yet another hierarchical officious oracle"). Thus the most widely used search tool and (with Amazon.com) one of the Internet's two best-known brands, was born, and the Internet was ready for the future.

Innovation Takes Time

· · · · ·

In the final decade of the twentieth century, the communications revolution took a huge step forward. The Internet, after the invention of the World Wide Web, has attracted more users in more countries more quickly than any other communications technology. It took thirteen years for television to reach a global audience of fifty million[32], thirty-eight years for radio, seventy-four years for the telephone – but only four years for the Web to achieve the same.[33] Traffic on the Internet now doubles at least every hundred days.[34] Even so, some of the novelties associated with the Internet have taken longer to catch on than the hype surrounding them might imply. Internet telephony, introduced in 1995, accounted for only 1 percent of global telephone traffic by 1999. Even Amazon took five years to account for 4 percent of American book sales.[35] [Fig 2-3]

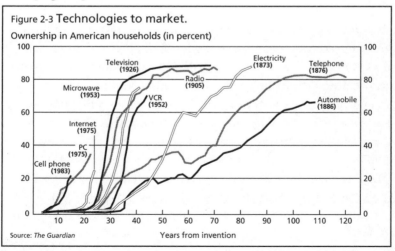

Figure 2-3 **Technologies to market.**

Ownership in American households (in percent)

Source: *The Guardian*   Years from invention

. . . . .

New inventions, even today, take a while to win converts. Millions of people must decide that the novelty is a sufficient improvement over what they had before to be worth the investment in learning how to use it. Even where companies are the adopters, progress may be hesitant. A study carried out in Britain examined the rate at which companies adopted numerically controlled machine tools. Large firms in growing industries began to use them in 1968; some small firms in declining industries had still not introduced them more than a decade later.[36] Diffusion of any new product follows an S-shaped path through the economy – slow-quick-slow. In other words, it is adopted slowly at first, then by a rapidly increasing number of people or companies, and finally by a few laggards. The Internet is still on the first slow curve of the S-path: only 6 percent of the world's population is on-line, and even in the rich world, the figure is only 35 percent. As of 2000, a mere third of American manufacturers used the Internet for procurement or sales.[37]

Because the price of a new technology tends to fall, and its reliability to improve, many people will wait before buying. The more complicated the novelty is to use, or the smaller the relative improvement over existing technology that it offers, the longer the delay is likely to be. Besides, innovation tends to be embodied in new investment, which occurs mainly during periods of robust economic growth. Not many people buy a new car merely to acquire global-positioning navigational equipment – but, when lots of people feel prosperous enough to trade in their old crock for something better, the new fleet will come with lots of novel gizmos on board.

So technological advance tends to happen most quickly in industries and countries that are expanding fastest. That is important for the communications revolution. Only in the 1990s did investment in computers rise steeply as a share of all investment in the American economy. In most other rich countries, it lags well behind America's levels.

In addition, inventions acquire new uses, often quite different from those with which they began. A classic example is the steam engine. Invented in the eighteenth century to pump out flooded mines, it was gradually applied to other stationary uses, such as milling. From 1820 to 1870 its main new use was in transport. Later it was employed to generate electricity, and therefore still produces most of the world's electrical power. It took more than a century for all of these applications to evolve.

The story of electricity is similar. Fifty years elapsed between the development of the technical capacity to generate electricity and the building of the first power station in the United States, in 1882. Fifty more elapsed before electricity was providing the power for four-fifths of American businesses and homes. As Paul David, an economic historian at Oxford University, argued in a famous paper, American industry took forty years to reorganize production in ways that exploited the electric motor efficiently.[38] It needed not just an understanding that the main gains would come from rearranging machinery so that it no longer stood beside the power source, as the use of steam required. To produce the world-changing innovation that was mass production and mass consumption also required the legal framework for large corporations and, arguably, universal primary education to create an industrial work force.[39]

In the history of past inventions, then, there are road maps that can suggest where today's period of astonishing innovation might lead. These teach the need for patience. The full impact of the changes in information technology and communications will not be clear for one or two generations, even if many of the effects materialize more quickly than they did during earlier bouts of technological change. In addition, pricing matters. Not only do sharp changes in relative prices convey powerful signals to individuals and companies, but patterns of pricing influence the way new technologies are used. The lesson of history is that the interaction of technology and economics is far more powerful than either force alone.

Sitting at your desk at home in the evening, you can, if you choose, watch CNN news on your personal computer, jerkily delivered to your home through a telephone line. Three technologies are thus united – the telephone, the television, and the networked computer. Among the many uncertainties of the next few years is how they will interact. Now that voice, video, and data can travel seamlessly along one pipe to a single device, how much convergence will people really require?

Separately, both the telephone and the television, two enormously successful and well established technologies, have been undergoing a revolution. Each is becoming more versatile and diverse. Each is finding different ways to charge users. And each is fighting to be the gateway through which most people enter the Internet. These changes are just as important as the more exotic development of the Internet.

Indeed, the telephone's mobile revolution has been even more successful, in most parts of the world, than the arrival of the Internet. In 1990 there were just over eleven million mobile telephones worldwide. In 2000, there were 650 million, compared with 500 million personal computers.[1] Every year since 1996, more people have subscribed to cellular telephones than to fixed ones and the gap is widening. By 2004, and quite possibly sooner, it is likely that one billion people – one person in six on the planet – will have mobile telephones, and the device will be poised to overtake fixed-line telephones. [Fig 3-1]

Here too is scope for convergence, for the combination of the mobile device and the Internet. Mobility brings with it two uniquely powerful qualities: the device is always with you, and it always knows where you are. Yet here again, companies must make huge bets on the uncertain

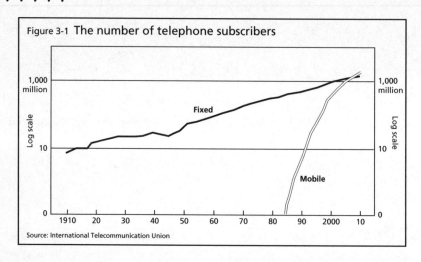

Figure 3-1 **The number of telephone subscribers**

Source: International Telecommunication Union

whims of millions of individuals. In 2000 five companies bet a staggering £22.5 billion ($35 billion) to buy the right to use radio spectrum in
Britain for "third-generation" mobile-telephone services that will combine voice, video, and data in new ways. Yet only in the special circumstances of Japan, where NTT DoCoMo sold thirteen million units of its
on-line i-mode service in the first two years after its launch, has convergence of this sort yet been a demonstrable success.

How convergence develops – how customers and services are eventually connected – will affect the future of many companies. Many of the
giant mergers of the late 1990s were driven by a desire to "own" the
customer: to control the path from person to product and back, extracting revenue and vital marketing information. Telephone operators, for
instance, have always "owned" their customers in one sense. But once
customers pass through a multitude of gateways on their way to what
they want – a service provider, a search machine, or a Web site – that
changes. Ownership (and the corresponding ability to extract profit)
becomes diluted.

Turn this problem over and you have another: how do customers pay
for their access to electronically available services? With television, they
have paid mainly through advertising or subscription; with the telephone, by subscription and a charge for use. The Internet, say some, is
free, but a moment's thought should dispose of that idea. Arcane
though charging may seem, it will determine the shape of the electronic future.

The Telephone

· · · · ·

During its existence of a century and a bit, the telephone has already transformed social and economic life. It has brought companionship, employment, and information to millions. Now the telephone network is changing. It is increasingly wireless and mobile rather than wired and fixed, and adapted to carry data ("packet-switched") rather than voice calls ("circuit-switched"). As a result, the cost of providing a long-distance call is collapsing, many parts of the world are acquiring a telephone service for the first time, and the telephone is increasingly portable and personal rather than stationary and communal. In addition, the telephone has become one of the main gateways to a new world: the Internet. Driving these changes is the novelty of competition, to which many telephone operators were exposed for the first time only in the closing years of the twentieth century.

The telephone business itself remains highly concentrated, but it is also, like any other fast-changing industry, full of newcomers, including newcomers from other countries and other industries. They are reshaping the services the telephone provides, for it still has enormous under-developed potential. Indeed, despite the growth of the Internet, what happens to the telephone may have more influence than the Internet on people's lives through at least the first decade of this century. Not only is the telephone roughly three times more common, even in developed countries, than the personal computer: it can also be used to check a bank account, take part in a radio show, sell insurance, lobby a politician, chat up a porn queen, check messages, or call a friend. Now the cost, flexibility, and range of this wonderful invention are being transformed, along with the lives of those who use it. [Fig 3-2]

The disappearing distance premium

In almost every country, ordinary telephone customers have always paid much more for a long-distance or international call than it costs the telephone company to carry it. This is changing. Eventually, no extra charge for duration or distance will be made for most telephone calls.

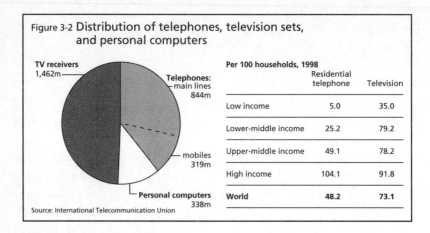

Figure 3-2 **Distribution of telephones, television sets, and personal computers**

TV receivers 1,462m

Telephones: main lines 844m

mobiles 319m

Personal computers 338m

Source: International Telecommunication Union

Per 100 households, 1998

	Residential telephone	Television
Low income	5.0	35.0
Lower-middle income	25.2	79.2
Upper-middle income	49.1	78.2
High income	104.1	91.8
World	48.2	73.1

The premiums charged for distance once partly reflected the fact that capacity was most limited on the international and long-distance parts of telephone networks. In addition, the rich were more likely to call long-distance than ordinary mortals and were therefore expected to cross-subsidize the price of local calls. Once telephone operators began to lay fiber-optic cables, shortage quickly gave way to surplus. These oil pipelines of the information economy are being laid at a tremendous rate: at least fifty now snake their way under the world's oceans. They have been laid most lavishly under the Atlantic and Pacific, and along busy routes in the United States.

Long-distance capacity is racing ahead, much of it built by relatively young companies such as Global Crossing and WorldCom to carry the surge in data traffic generated by the Internet. Total capacity on fiber-optic cables laid under the world's oceans will grow more than forty-fold between 1999 and 2002, to nearly 12,000 billion bits per second: enough for every American to chatter simultaneously.[2] At the same time, capacity on the other main long-distance technology, the satellite, is also growing fast. Building capacity now costs much less than it once did, mainly because each new cable or satellite has more capacity than ever before. The cost per year of constructing one cable transatlantic voice path – the capacity to carry one telephone conversation – is now less than $10.[3] And the figure will fall even further as better multiplexing techniques squeeze more channels on to the same fiber at different frequencies. The basic cost of carrying an additional telephone call across the Atlantic is thus as near to zero as makes no difference. [Fig 3-3]

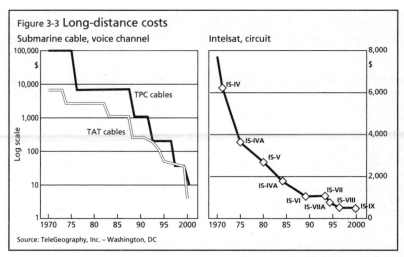

Figure 3-3 **Long-distance costs**

Source: TeleGeography, Inc. – Washington, DC

Of course, there are other costs. The costs of the local loop – the last part of a telephone network, which connects small businesses and domestic customers to the main system – have fallen only slowly. In addition, telephone companies tend to employ lots of people for activities such as billing and marketing. Overheads cost money. But the average price of an international call has already fallen from $2 a minute in 1990 to fifty-three cents by 2000. By 2005, estimates Michael Minges, an economist at the International Telecommunication Union (ITU), it will be down to 17 cents. In each of the last two years of the twentieth century, the average price dropped by around 20 percent.

Increasingly, the world's long-distance routes carry data rather than voice. Early in the twenty-first century, indeed, data will overtake voice. The demand for bandwidth is driven by the growth of the Internet, but it also responds more to changes in price than do voice calls. Each time the price of capacity drops by half, some Internet service providers double their purchases.[4] The death of the distance premium thus drives the Internet even more vigorously than it does other kinds of communication.

The impact of government

It is not just technology that has driven down the price of international calls. Government agreements have accelerated the decline. Until 1998, the right to carry international calls into or out of a country was generally a monopoly. In that year, thanks partly to an agreement negotiated at the World Trade Organisation, countries began to liberalize telecommunications.

In addition, the international accounting system began to break down. This arrangement split the cost of a call between countries. If one country put through more calls to another than it received, it handed over a settlement payment to even things up. In the mid-1990s the United States, annoyed to find that it paid out much more than it received under this arrangement, began unilaterally to set benchmarks intended to drive those rates closer to true costs.

These changes have almost ended the premium for distance on certain routes, like the Atlantic, where competition is intense. A portent of the future may be a Californian company called bigredwire which, in July 2000, offered free long-distance and international calls as an incentive to people willing to use its entertainment Web site. Another is the creation of a single telephone code for Europe, intended to allow call centers to give their customers one point of contact, whichever country they are in. Europe will take time to agree how to price such a service, but the promise of a borderless continent is there.

Declining rates will have powerful effects. As the ITU puts it, "It seems inevitable that a world without distance is also a world without borders."[5] Not only do some countries feel that their national identity is bound up with their communications monopolies – national post, national telephone giant, and so on. They are also, more importantly, used to the idea that calls should cost less within their borders than beyond them. The difference has helped to define "abroad" and "at home."

Thus in Norway, with three million people in a country more than a thousand miles from end to end, long-distance telephone rates have been kept low for many years in a deliberate attempt to foster a sense of nationhood. In the European Union, by contrast, telephone tariffs have long been divisive rather than harmonizing. During the working day it can cost more than three times as much to call Bonn from London as it does to call the slightly more distant city of Edinburgh. Soon, such divisions will be gone. In the rich world, at least, local and international calls to and from fixed lines will cost the same. The death of distance in telephone pricing may not make people love their neighbors, but at least they would be more likely to talk to one another.

A new pattern of pricing

Competition raced through the telephone markets of the rich world in the final years of the 1990s, bringing with it radically different pricing policies. By the beginning of 1999, almost all the markets of the OECD were open to unrestricted competition. Poorer countries moved more slowly: in the world as a whole, only a quarter of the market for basic telephone services had opened to competition by then.[6] World markets for mobile communications and Internet provision were freer than those for fixed lines, however. Small wonder that these were by far the fastest growing sectors.

Several pricing trends are emerging. The distance premium no longer cross-subsidizes local tariffs. In the past, the profits from long-distance and international business often helped telephone companies to hold down their charges for local calls. In the United States, until the late 1990s, access charges mandated by the government and paid by long-distance carriers to local ones represented more than a third of the price of a long-distance call. As such cross-subsidies vanish, local call charges rise.

Another trend is a shift from charging per call to charging for the basic use of the network. Outside the United States and a handful of other countries, most telephone companies bill customers mainly by how much they use the network. Increasingly, access charges – an up-front payment for the use of the network – will become the main tariff.

Finally, the share of calls paid for by the caller will diminish. As individual calls become less expensive, companies will find it attractive to offer toll-free services to their clients and suppliers. In the United States, toll-free calls already account for a hundred million calls per day, and AT&T reckons that 40 percent of its domestic traffic travels to toll-free numbers.

So far, the overall pattern has been for pricing to become more fluid and flexible. Just as airlines calculated that selling unused seats, however cheaply, is better than flying with those seats empty, so telephone companies try to fill their networks by varying rates according to demand. As a result, just as no two passengers on a flight seem to pay the same price for their tickets, so, increasingly, no two callers pay the same price for a call. Users all get the same basic service (they all get on the same plane, or call the same town), but they pay according to a range of special considerations, of which quality of service tends to be the most important.

In the United States, this trend is already well established. The actual billed prices people pay for telephone calls are less than half the published price of a peak-rate call.[7] A newer trend, which may supersede complexity, is starting to emerge. For domestic callers in particular, tariff structures are becoming simpler. One of AT&T's most popular plans offers uniform rates, with no variation for time of day or distance within the United States.

Most important of all the pricing conundrums facing the telephone companies, though, is how to deal with the two big new business opportunities that sprang up in the second half of the 1990s: mobile telephony and the Internet. The cellphone has now overtaken international calls as the main source of profit for many telephone companies. The distance premium has dwindled, but the mobility premium has not. It can, for example, cost six times as much to make a local call in Geneva on a mobile telephone as to make a transatlantic call from the same city. If the call from the mobile is to another mobile on a different operator's network, the premium may rise to eight times as much.[8]

The future of that premium will be influenced by the efforts of the telephone companies to bring mobility to the Internet. One of the biggest advantages of fixed-line telephones over mobile ones is that they got to the Internet first, and can offer lots more bandwidth. Because the telephone companies' charging arrangements are so crucial to the way Internet use spreads, they are discussed later in this chapter.

Wiring the world

While telephone companies in the rich world struggle to think of new and more lucrative services to offer their sated customers, two out of three people on the planet still have no access to a telephone at all. But that is changing fast. And, in building their telecommunications networks, poor countries have two advantages: many new technologies, such as wireless, cut installation costs, and a new network will be state-of-the-art. Many developing countries, including Cambodia and Bolivia, have fully digital networks. Good services require vigorous competition and hefty investment, however. Thus some developing countries are wisely allowing foreign firms to roll out part of their networks. Australia's Telstra has installed pay telephones in Phnom Penh and Deutsche Telekom is building a fixed-link network in Indonesia.

Such policies make sense not only because they bring people inexpensive telephone services, but also for an even more important reason. Good telephone services are especially important for poor countries because they bind the most impoverished, distant regions into the rest of the country, and allow these nations to start competing on the same footing as rich rivals.

Connecting rural areas. Proportionately more people live in the countryside in poor countries than in rich ones, and they are less likely to have access to a telephone. In Bangladesh, for instance, around 80 percent of the population live in villages – but 80 percent of the country's telephone lines are in its four largest cities.[9] Not only do villagers have less political clout than city dwellers; it also costs more to install a fixed line to a village.

The answer lies in wireless: either mobile telephones or a technology known as "wireless local loop," which involves installing a small fixed radio antenna in a home or shop to receive calls from a nearby transmitter. While a mobile telephone uses quite a lot of its spectrum to switch from one transmitter to the next, a fixed antenna can be permanently tuned to the correct base station.

The benefits of good communications in the country are even greater than they are in towns. They can bring news, education, medical and agricultural advice, and link farmers directly to markets, enabling them to check on the prices traders offer for their crops. Rural telephones are also a good investment. They frequently earn larger revenues than those in the city because country-dwellers make – and attract – many more long-distance calls. In fact, it is just as important to connect cities to their rural hinterlands as to link cities to the rest of the world. Failure to do so will mean that the impact of the death of distance on poor countries will be divisive, not unifying.

Improving competitiveness. One of the greatest effects of investing in telecommunications will be to reduce the gap between developed and developing countries. Since about 1990, international-call charges have been converging as developing countries cut tariffs faster than rich countries (by roughly 3 percent a year compared with the rich countries' 2 percent).

The more industrialized developing countries are now on track to be this century's communications powerhouses. Some are beginning to offer a low-cost location for telecommunications services, just as they have long provided a low-cost location for the manufacture of textiles and electronic goods. Call centers serving the rich world have sprung up in countries as diverse as Panama and the Philippines, for example.

Add the Internet to the telephone, and people in poor countries acquire even greater benefits. Private enterprise is connecting PCs to terminals in all sorts of surprising places. In the poorest developing countries, indeed, the number of new Internet host computers has grown faster than fixed-line or even mobile telephone connections. Thus in India, entrepreneurs convert village telephone booths into makeshift Internet cafés, running the power off a diesel generator to guard against power cuts. These cafés are giving villagers more power not only to ask a better price for their crops, but to deal with corrupt local officials. At present, Indian villagers pay hefty bribes to local officials to see land records. In future, the Internet will enable them to see such information instantly and to discover when records have been altered.[10]

Enormous gains will flow from plugging developing countries into the world's telephone network. Services and information that have long been taken for granted in the rich world will suddenly become accessible. Information brings power, education, health, and wealth. There could be few better investments.

Wireless and mobility

Among Scandinavian men in their twenties, it has already happened.[11] Everywhere else, it will come within the first decade or so of the century: the moment when ownership of mobile telephones begins to approach 100 percent and people stop talking about a "mobile" or "cellular" telephone because most voice calls are made on portable devices. Fixed-line telephones will be used when the caller wants to reach an entity – a company, say, or a household. They may also, because of their lower charges, be preferred for lengthy or international calls.

Wireless connections change the telephone in several ways: they are relatively inexpensive to install and maintain and so can be used (as in the developing world) to provide inexpensive links to fixed telephones. They can carry broadband signals, and so create fast Internet access in rich countries where fixed-line telephone companies have been slow to do so. But by far the most widespread and popular use of wireless telephony has been to provide mobility, and it is this that has revolutionized the telephone. In addition, the mobile telephone is a personal, not communal device. That changes the nature of telephone ownership.

As the 1990s drew to a close, the quality of service that mobile telephones offered was racing forward. Most countries switched from analogue to digital networks, which can carry three to six times as much traffic and offer greater security, better reception, and features such as caller identification, call forwarding, and three-way calling. The size of the telephone shrank, thanks to better battery and radio technology, while beautiful design, especially by Nokia, turned mobile telephones into tiny and covetable objects with several days of standby time.

The huge popularity of mobile telephones also shows the value of competition. In most countries, the main competition to the fixed telephone comes from the mobile network. Many governments that are reluctant to allow competition with their own fixed-wire monopoly have still allowed two or more mobile operators to set up. Everywhere, the result has been falling prices, innovative services, and a steep rise in use.

In the near future, mobility will drive three other important trends. It will spread telephone ownership to new social groups and new parts of the world; it will transform access to the Internet; and it will provide a link with fixed and moving objects, changing the nature of many industries.

New competition. The price of mobile calls has fallen rapidly all over the world. But the fastest falls have come in countries where two or more services compete. When Chile allowed in new operators in 1998, prices promptly dropped by 40 percent and the number of new customers doubled. Much the same happens in other parts of the world. In Lebanon, where competition thrives, mobile rates are the lowest in the region: seven cents a minute, compared with a regional average of forty to fifty cents.

Bring in new entrants, and the market immediately gets a fillip. In the late 1990s, many of those rich countries of the OECD that had previously had only two mobile operators acquired three or more. Almost invariably, ownership of mobile telephones promptly jumped, relative to other countries.[12] Charges came down and new products, especially pre-paid cards, boosted use in new markets.

In addition, competition from mobile companies influences the fixed network. When, in 1995, the government of Ghana licensed a mobile operator, the fixed monopoly dropped its prices by half and started operating outside Accra.[13] Tariffs change, too. Mobile-telephone tariffs are usually the same throughout a country. As ownership of mobile telephones grows and mobile tariffs fall, fixed-line companies are forced to fight back. Norway and Iceland now offer a single national rate for calls on the fixed network.[14] In Finland, where two-thirds of the population have mobile connections (and jabber away on them even when sprawling naked in the sauna), all calls on the fixed network pay one of two tariffs, one regional, one national.

Wider access: In the rich world. Fierce competition for customers has encouraged mobile-telephone companies to dream up new ways to reach them. As a result, there is less of a "digital divide" in mobile-telephone ownership than in Internet access. A survey in Britain by Oftel, the telecommunications regulator, found that 70 percent of the highest-income groups had mobiles, compared with 39 percent in the lowest group. It was the old, rather than the poor, who were most likely not to own a mobile telephone.[15]

The reason for the relatively egalitarian pattern of ownership is the mobile's most successful innovation: pre-paid cards, which enable people to buy telephone time as casually as they might pick up a newspaper or a packet of chewing gum. In many countries, pre-paid cards were the main reason people bought mobile telephones in the late 1990s. They spread like wildfire in countries such as Italy (where, in 1998, they accounted for 74 percent of mobile subscribers) that had previously been sluggish telephone users. Even in mobile-crazy Sweden, pre-paid cards had grabbed a quarter of the mobile market within two years of being introduced.[16] [Fig 3-4]

For telephone companies, the beauty of pre-paid cards is their simplicity: no credit vetting, no billing. In the United States, where about 30 percent of applicants for conventional cellphones are rejected

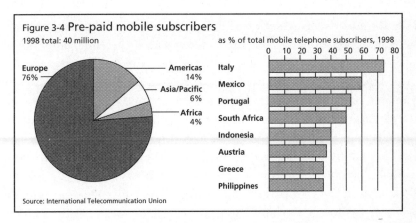

Figure 3-4 **Pre-paid mobile subscribers**

1998 total: 40 million

Europe 76%
Americas 14%
Asia/Pacific 6%
Africa 4%

as % of total mobile telephone subscribers, 1998

Italy
Mexico
Portugal
South Africa
Indonesia
Austria
Greece
Philippines

Source: International Telecommunication Union

because of poor credit rating, the idea has huge potential. And the telephone company can split the takings with the retailer, building in a big incentive for unlikely stores to sell the phones.

For customers, they have other virtues. In Italy, they are called "Mafia phones": no contract, no trace of the user.[17] For most users, the biggest selling point is that they know for sure what they will spend. This has made pre-paid cards useful for companies that employ staff who take many incoming calls, such as plumbers or security guards. They are even more useful to parents. BT Cellnet has a scheme that allows a card user to nominate one fixed-line number which can be called free of charge, so that, say, children can phone home free. Lots of card companies aim their products at the young by offering free calls in the evenings or at weekends to lure teenagers, or by decorating them with Winnie-the-Pooh, for example, to appeal to the youngest market. The pre-paid card has thus brought telephone ownership within the reach of the very poor and the very young.

Wider access: In the poor world. A wireless network is far less expensive to build and to run than a fixed one: no roads to dig up, no copper wire to lay (or to be stolen), no need to pass the properties of people who do not want to be connected, low maintenance costs. High overheads mean that the costs of fixed networks will never fall below a certain point. Wireless networks, by contrast, grow ever cheaper and more powerful.

So the technology has begun an even greater revolution in cash-strapped developing countries. In many of them, people wait for years to be connected to the fixed network. Now the wireless revolution is allowing whole nations to vault a stage in development. In China, to take the most spectacular example, telephones were rare luxuries not long ago. By the end of 2000, the country had fifty million subscribers, making it easily the largest market after the United States.[18]

In Finland, where many folk have second homes in the wastes of Lapland, one in five households has only a mobile telephone.[19] In some developing countries that situation will become even more common. More people will have mobile than fixed-line telephones – not because they are smart or convenient, but because they are the only telephone available without interminable delays. One example is Cambodia, where more than 70 percent of all telephone subscribers are mobile. Another is Morocco, where fixed lines cover 6 percent of the population, the mobile network 85 percent.[20] Some forty million people around the world are waiting for a telephone. Cellular will end those queues, then go on to put telephones in the hands of billions of people who never dreamed of owning one.

All the tools cellular companies use to reach the poor and the young – pre-paid cards, ingenious tariff schemes aimed at particular markets, sales through non-traditional outlets – help them in developing countries too. The first network using only pre-paid phones started in Senegal in 1999. In Bangladesh, a new breed of entrepreneur has sprung up: "phone ladies," who make their living by buying cellphones with loans from the Grameen Bank, a private company that more usually makes tiny loans to villagers to buy cows.[21]

When the Internet spreads through countries such as Cambodia, Morocco, and even China, it will not arrive on the PC, as it has so far done in the United States and Europe. It will arrive mainly via a new variant of the mobile telephone.

Paying for mobility. In "The World in Your Pocket," an article published in *The Economist* describing the spread of mobile telephones, Adrian Wooldridge recounts the curious tale of Ann Schrader, a reporter with the *Denver Post* at the time of the tragic shootings at Columbine High School in Colorado in 1999. As news began to come into the office, Schrader's boss despatched her to a local hospital to see the damage. She paused to grab a mobile phone on her way out, but the paper had only half a dozen of them and they had all been booked out.

When she was half way to the hospital, her pager went off. She searched for a telephone to call her office, and eventually found a branch of 7-Eleven. Her office redirected her to the casualty reception near the school. On her way there, the pager went off again; again she scurried for a telephone, finding one at last in a liquor store. When she reached the school, she managed to borrow a mobile telephone – only to find that so many people were trying to make calls that she could not get a line.

"There are several surprising things about this story," says Wooldridge. "It is surprising that a reporter on a reputable newspaper should not own a cellphone, or at least be able to put her hands on one. It is surprising that a middle-class westerner in a professional job should rely on a crude piece of technology – the pager – that is being abandoned by humble laborers in Hong Kong. And it is surprising that the wireless operators in a big city should not be able to handle the surge in traffic after a disaster."

In fact, the last point is the least surprising: even the wired network sometimes has to block calls to an area where a disaster has taken place, because the network has insufficient capacity.[22] More striking has been America's relatively slow uptake of mobile telephones. At the end of 1999, the United States was not even in the top-twenty league of countries for mobile penetration. The proportion of Americans with cellphones, at 31 percent, was lower than that of the Japanese, with 45 percent, and less than half that of Finns, with 67 percent.[23] Americans have been slower than Europeans and Asians to switch from analogue to digital, and so slower to acquire the fancy services that people in other countries adore, such as short messaging and call-forwarding. Pre-paid cards, which accounted for so much market growth in Europe in the late 1990s, have also spread more slowly. [Fig 3-5]

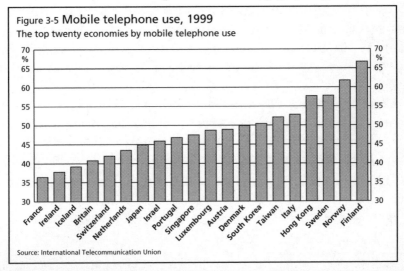

Figure 3-5 **Mobile telephone use, 1999**
The top twenty economies by mobile telephone use

Source: International Telecommunication Union

Why have mobile telephones been so much more successful in other parts of the world? Partly, because America has not agreed a single national standard for digital telephones on the lines of the GSM standard, which enables Europeans or Asians to use the same handset everywhere from Australia to China to Italy to Iceland. Partly, because America's system of regional licenses for spectrum creates the need for a huge secondary industry to handle "roaming" between operators, something that users in other countries do only when they go abroad. But mostly because the system of charging requires the recipient of a call to foot part of the cost. Everywhere else, the caller pays all of the cost (except when a call crosses national borders). To avoid unexpected charges, Americans habitually use pagers: you leave a message on the pager, and they call you back on the mobile. A study in 1999 by the Yankee Group found that more than 30 percent of Americans with cellphones never left their phones switched on, and more than 20 percent never left them on for more than two hours.

It will be hard for the United States to switch to a "calling-party-pays" system. The key reason is the American telephone-numbering plan. Callers in Europe or Asia know when they are dialling a mobile telephone because the number is different. (In Britain, for example, all mobile numbers begin with 07.) In the United States, many mobile numbers have area codes that make them indistinguishable from landlines. So the Federal Communications Commission will allow a switch in the system of charging only if a recorded message tells callers that the number they are ringing is a cellphone. That would be a considerable deterrent.

In addition, mobile calls are often billed by the minute, whereas local calls in the United States are generally charged on an "all-you-can-eat" basis. So the charging system that has been so encouraging to the Internet (see below) simultaneously helps to discourage the use of mobile phones.

Short messages. When mobile telephones first appeared, hardly anyone foresaw their amazing success. That was equally true of the short message services that enable people to use their mobiles as pagers (but pagers that can also connect with the Internet) and exchange messages of up to 160 characters each. In April 1999 users in Europe sent more than one billion SMS messages, and some telephone companies said the number was up by 800 percent on the previous year.[24]

Many of the users are youngsters, using a service that is cheap and fun and does not – or not necessarily – make a noise in the library or need to be audible above the din in a bar. Businesses are also finding uses. Offices can send pick-up information and delivery addresses to workers on the road, for instance. Companies can track freight loads and plan delivery routes by sending global-positioning system (GPS) information, together with a vehicle's registration details and average speed.

From SMS it is a short step to other mobile data services, some serious, some not: mobile banking (in Finland), flight information, weather and parcel tracking (offered from an American Web site called hz.com), down-loading tunes to use as ringing tones (Finland again, with a dismaying propensity to choose themes from South Park), or blind dating (Hong Kong). It is not clear, however, whether the success of SMS suggests that there is a huge unmet demand for mobile Internet services – or that what most folk want is a mobile data service that is simple, cheap, and more like e-mail than the World Wide Web.

Connecting to the Internet. Merging the mobile telephone with the Internet became in the late 1990s the Holy Grail of the communications industry. Around the world, companies were racing to develop third-generation (3G) mobile devices that would bring video to tiny mobile screens. All sorts of companies with mobile gadgets on the market are approaching the challenge from different directions. One example: 3Com, the Californian company that makes the PalmPilot personal organizer, is adding some Internet services to a new mobile device.

As of 2000, though, the main commercial success in bringing data to the mobile telephone was NTT DoCoMo's i-mode. Slung around the neck of young Japanese, it allows them to send e-mail, download jokes, and buy tickets for a movie. Soon it will add GPS features that will beep when a friend is near by, or when its wearer passes a café offering a hot deal on cappuccino. I-mode offers a huge array of specially designed services, almost all in Japanese. It does not, however, offer access to the Internet at large. Its success seems to spring partly from the fact that it is always on – users do not have to turn it on, as they do a PC – and partly from the fact that it does not charge by the minute. The extremely high price of access to the Internet through a conventional telephone probably also has something to do with it. If local calls in Japan were cheaper, the i-mode service might be less attractive: it operates at a modest 9.6 kilobits per second, giving about the same experience as an Internet connection in pre-Web days.

As new mobile Internet services develop, with higher bandwidth and lower prices, many more people will find their way on to the Internet. But they will, at least initially, use it differently, seeing it as an addition to the capability of their cellphones rather than an alternative to the PC that many of them will never own. So they may want to bet (and Hong Kong's Jockey Club already allows people to place bets over its private network using wireless devices); they may want to trade shares; or they may want to bid in an auction. Most useful of all could be the smart phone as credit card: point it at a petrol pump or soft-drinks dispenser, and it will authorize the purchase of the right kind of liquid and add the cost to your telephone bill. Here, the strength of the telephone companies will lie in their established billing mechanisms, tailor-made to record bills for tiny amounts of money. But all of these uses share a common characteristic: they require good security even more than plentiful bandwidth.

Connecting with objects. It is not just people who are constantly on the move. The most pervasive use of wireless in the future may be to monitor things, rather than to connect human beings. Link wireless with the Internet, and such telemetry can become vastly more sophisticated. It is becoming rapidly less expensive to build such links thanks to innovations such as Bluetooth, which uses shortwave radio frequencies to link together devices such as cameras, printers, and laptop computers.

Anything that moves can be tracked or connected: pets, criminals, grannies with Alzheimer's. Cars and trucks offer the biggest opportunities of all: to receive traffic information and text-to-speech e-mail, and to transmit data to sales departments and fleet operators. General Motors' advertisements boast that its cars are fitted with the On Star safety feature, which signals the rescue services if a car runs into trouble. Some companies are developing ways to tune an engine remotely or to link cars with parking spaces. Volvo's racing cars already carry wireless gadgets to tell mechanics waiting at the pit stop when they will arrive and what will need to be done. One day, such devices might improve maintenance for everybody's cars.

Other applications include:
- Inventory control: monitoring the position of electronically tagged containers, or checking stock on supermarket shelves

- Navigation: electronic maps pinpoint the location of a truck or tell a farmer which parts of a field to spray

- Alerting: tags track stolen motorcycles or monitor the state of a patient's heart to warn of an impending attack

- Toll payments: billing the correct fee for vehicles passing through an electronic checkpoint or driving down congested streets.

All told, NTT DoCoMo estimates that, by 2010, only a third of its 360 million customers will be people. The rest will be cars (one hundred million), bicycles (sixty million), portable PCs (fifty million), boats, and other objects. All will therefore be easier for their manufacturers to monitor. They will also be harder to steal. Indeed, apart from the digital camera, it is hard to think of a technology that will make a more dramatic difference to reducing crime than wireless (give or take a few Mafia phones).

The future of the telephone

It is easy to forget what a remarkable technology the telephone is. A single standard allows almost one billion machines to connect with one another, carrying speech, data, and faxes to compatible machines from Michigan to Morocco. And the telephone keypad – with ten numerals, the alphabet, and a few extra symbols – can be used, at one extreme to ring the house next door and, at the other, to turn the telephone into a miniature PC.

For a familiar object that has been in use for more than a century, the telephone shows an extraordinary ability to evolve. It is mobile and interactive, in the sense that a user can send a return signal by pressing a key. As communications develop, so the telephone's ubiquity and simplicity will ensure that it develops many more uses.

Apart from offering a gateway to the Internet, the telephone will remain the principal electronic device that individuals use to communicate with one another for at least the first two decades of the new century. It will become ever more personal, as telephone numbers eventually become allotted to individual users rather than to individual machines. As a result, the personal number may become a reliable way to reach someone even after five or ten years, even when their jobs and home addresses have altered. It will become part of a person's identity, like a social-security number or date of birth. Or perhaps personal numbers will be replaced by something more memorable and human than a string of digits.

The telephone's ability to carry the human voice, uncomplicated by pictures, is part of its charm and intimate power. A telephone conversation is an experience unlike any other that sighted people normally engage in. Its one-dimensional quality means that, like talkers in the dark, people on the telephone may feel able to say things to each other that they would find difficult to say face to face: to sell, to threaten, to seduce, to lie. Invisibility liberates. You can chat to your boss or your mother while glancing through the newspaper or getting dressed. But the camouflage is only partial. Undistracted by body language, which may deceive the eye, the ear often picks up faint emotional signals – distress, glee, equivocation. The telephone sharpens the senses, allowing the ear to spy on the voice.

So, while the telephone's functions will become more intertwined with those of the Internet, there will remain a demand for that most satisfying of all forms of electronic communication: one human being, speaking to another, unseen and far away, carrying ideas and emotions across distance.

The Television

. . . . .

Even the *New York Times* sometimes makes mistakes. "Television," predicted that great newspaper on the occasion of the 1939 World's Fair, "will never be a serious competitor for radio because people must sit and keep their eyes glued on a screen; the average American family hasn't time for it."[25]

Midway through any given evening in Europe or the United States, about half the people are doing the same thing: watching television. This activity now takes up more of the average person's time than anything else, apart from sleeping and working. People in Western countries typically spend between one-half and one-third of their leisure time in front of the box. The ability of television to entertain, inform, titillate, and lure made it not only the chief leisure activity of the twentieth century but also the most influential cultural invention since the printing press.

At the end of the second world war, a mere eight thousand homes worldwide had a television set. By the close of the twentieth century, there were about one and a half million sets and more than two thirds of the world's households had television. Yet the basic technology of the set barely changed during the intervening fifty years.

Now, though, technology is transforming television as surely as it is the telephone. Two enormous changes are under way: a rise in the capacity of delivery systems, so that people can receive not tens but hundreds of different channels, and a fall in the cost of making and distributing programs. These changes will alter the nature of television. A third change is just beginning, and may yet harbor surprises: the convergence of television with the Internet.

The changes will differ in important ways from those that are bringing the death of distance to the telephone. First, most of what people watch is "free," so the fall in delivery costs that is now cutting the price of a telephone call will not cut the price of receiving a television program. Instead, ironically, people are being asked to pay directly for the programs they most want to watch.

Second, in many wealthy countries, people already seem to be watching as much television as they want to. Indeed, in some countries, viewing time is declining. The television is having to compete for viewers with other uses of the screen, such as playing electronic games and surfing the Internet. So, while time on the telephone (another competitor to television) is rising, time watching the box will fall. Television will offer more to watch, but viewers will have less time to watch it.

In developing countries, however, where channel choice may be limited and transmission hours short, both viewing hours and levels of television ownership are far below those in the rich world. The rise in set ownership and the spread of multichannel television represents a vast new market. Some of it will be fed by the exports of the rich world; much more will come from new local entertainment industries.

As channels multiply, the most basic question is whether television becomes more fragmented. Television has been a uniquely unifying national, and indeed global, phenomenon. Never before have so many people held in common a core of shared cultural experiences, whether it be viewing *Friends* or second-guessing the outcome of the O. J. Simpson trial or weeping over Princess Diana's funeral. That shared experience constitutes a durable communal bond. You may not know the names of your next-door neighbors but you can be fairly sure that, over the past few days, they have watched some of the same programs as you.

Now, with the vast expansion of programing, everyone will be able to watch something different – although that is not necessarily what people really want to do. The "five hundred-channel universe" that John Malone, chief executive of American cable giant TCI, acclaimed in 1992 seems at last to be imminent. The television will become a personal piece of equipment – as it is already in the United States, where there are four receivers for every five inhabitants – and so turn into "Me TV," more like a mobile telephone than a communal source of entertainment.

Undoubtedly, people will have more viewing choice. That will be revolutionary, especially in those large tracts of the developing world where television consists of a few mind-numbingly boring and amateurish state-controlled channels.[26] Even in Europe, choice is relatively recent. Austria, admittedly an extreme example, drafted a law allowing private television stations only in February 1997.

Viewers everywhere will one day be able to pick their programs in a global market. They may still choose to watch local fare: imports, with the partial exception of American programs, tend to be less popular than national shows. But, armed with a credit card and a remote control, people will eventually be able to order programs from anywhere they choose. The television business will then become as global as the music business. So, perhaps, will the cultural values it instils.

Three key changes will take place. First, distribution networks will switch from analogue to digital and new delivery systems will develop. Second, viewers will increasingly pay directly for content. And third, the Internet will enhance and extend the role of television in some ways and deplete it in others.

Changing distribution

Two changes are taking place at once. First, the old clique of over-the-air broadcast networks is being challenged by newer delivery systems, such as cable and direct satellite broadcasting. Second, all delivery systems are becoming digital, hugely increasing their capacity. So a medium that has long been defined by its limited capacity – a handful of analogue department-store channels, all showing a jumble of different programs to vast audiences – is now acquiring more or less unlimited capacity.

More ways to distribute. In most of the world, the only sort of television available for most of the past half century has been broadcast over the air. But spectrum is limited, and analogue television needs lots of it. So the predominance of over-the-air broadcasting has effectively restricted most people in most countries to four or five channels, available nationally.

This restriction of choice has had enormously important consequences for the nature of television. First, because competition has been limited by spectrum shortage, television has been highly regulated, often even owned, by government. Second, channels have had an incentive to cater to the largest possible audience, making television the ultimate mass medium. And third, television's huge audiences have shaped the development of mass-market advertising, and so, for instance, the development of brands.

While the number of analogue over-the-air channels is tightly limited, most cable-television systems can deliver up to about fifty analogue channels.[27] In only a few countries, however, is cable television common. The United States has easily the world's most extensive cable network, passing 80 percent of the country's 115 million homes and connecting 64 percent of the homes passed.[28] Only in a few European countries – Belgium, the Netherlands, and Switzerland – does cable run past a higher proportion of homes, mainly because these three small countries are surrounded by larger neighbors with powerful transmitters that make broadcasting difficult. All told, the United States accounts for more than half of the rich world's cable-television sub-scribers.

Now, many countries have begun to build or upgrade their cable net-works. The fastest growth has been in developing countries, where cities with lots of high-rise buildings find cable a relatively inexpensive way to take television to lots of customers. China has more cable-tele-vision subscribers than any country except the United States, and India has more homes with cable television than with telephones.[29]

But a bigger revolution in delivering television has been the start of what Americans call "direct broadcasting by satellite," or DBS, and Europeans call "direct-to-home," or DTH. Initially, satellites were used for telecommunications. Then they were used to send programs to the cable companies, which would receive them at the cable "head end" and transmit them to subscribers. The big novelty of the 1990s was the development of satellite television delivered straight to the customer, who buys or rents a small receiving dish.

The fastest growth in DTH satellite in the 1990s took place in Britain, where BSkyB, a company part-owned by Rupert Murdoch's News International, grew to become at one point Europe's largest media group in terms of market capitalization. Lots of other satellite-televi-sion services followed: indeed, in the mid-1990s, the use of satellites for broadcasting was more widespread and growing faster than the use of satellites for telecommunications.

The boon of satellite delivery, as in the case of mobile telephones, another wireless technology, is its low start-up cost. Cable companies, like fixed-line telephone operators, have to pay all the initial costs of building their networks, and must run them past homes that do not want their product as well as past those that do. Direct-to-home broad-casters merely rent capacity on a satellite. In addition, while cable sub-scribers assume the cable company will foot the bill for connecting them, satellite subscribers usually buy the satellite dish and arrange its installation themselves.

The bonus for viewers of many delivery systems should be more choice: not just choice of content, but also, more importantly, choice in how to receive that content. Cable companies are generally local monopolies. Just as it is clearly desirable that telephone subscribers should be able to choose their company, so viewers should be able to choose from several companies that can deliver television to their homes.

Going digital. Digital television represents the first big change in the way a signal is delivered since the launch of color. Compression enables many more digitally encoded channels (already between four and sixteen, depending on the amount of movement) to be squeezed into the same space required for a single analogue channel. Through digital delivery, televisions can use many of the tricks that were once confined to computers: storing information or programs, for instance, and manipulating it in various ways.

Digital technology is being used for all three main delivery systems. Over-the-air broadcasting is the last to switch, but it brings a unique advantage: portability. A digital over-the-air set, unlike a cable or satellite set, can be carried around like a pager, a radio, or a portable telephone, and thus may one day be combined with one or more of these devices. (It also produces excellent reception. A wire coathanger, stuck in the aerial slot during one British demonstration, still produced a perfect picture.)

All the channels available on analogue services speedily appear on digital too. Some viewers have begun buying digital sets. Many more will be reluctant to do so. Digital television offers better picture quality, but the gain over analogue is greatest in the United States. In most countries, which use a different transmission standard, the quality difference is more like that between AM and FM radio than between black-and-white and color television. Just as mobile-telephone companies in most of the world coaxed customers to abandon their analogue phones for digital by offering wondrous new services and conveniences, so television companies will have to do the same.

In time, governments will force the pace by insisting that analogue broadcasters turn off their transmission and switch all their programing to digital. That will spark a debate about who should reap the main gains from switching. If the huge amount of spectrum that analogue broadcasters enjoy is their property, they have an immense incentive to pay their viewers to switch, by subsidizing the high cost of digital sets. But if spectrum is public property, why should its value go to the companies lucky enough to be using it? There will undoubtedly be plenty of furious political rows before analogue broadcasting shuts down.

· · · · ·

Changing content

The delivery revolution is leading to another change: in the content of television and the way people pay for it. The potential global audiences for really popular material are rising as more people have television and as viewers are offered a growing number of chances to see the same small selection of top films and programs. Television companies are getting better at squeezing from audiences every penny they are willing to pay in order to watch, say, an important sporting event or movie. Just as Hollywood now releases films in a carefully calibrated sequence – first to the cinema, then for video rental, then for video sale, and finally for screening on television – so, increasingly, a large proportion of television programs are packaged and repackaged for a succession of different "windows," each available at a lower price than the one before, to squeeze out maximum revenue. As a result, the control of exclusive rights to content, the raw material of television, becomes ever more valuable.

At the same time, one of the main sources of finance for television is under threat. New television technologies make it easier for viewers to skip the ads. As a result, television companies will need to rely more on subscriptions than they have done in the past.

Boutique television. The fragmentation of television audiences began in 1980, when Ted Turner launched a novelty: Cable News Network, offering nothing but news around the clock. America's broadcast television industry was snooty, but the idea caught on. Not only has CNN sliced up and repackaged the output from its newsroom into more and more different channels (one for sports news, one to be shown in airports, and so on), but other companies have done likewise with channels such as Viacom's Nickelodeon, aimed at children, and Disney/ABC's ESPN, aimed at sports fans.

Most new channels in the 1990s were of this sort, taking a radically different approach from the broadcast networks. Not only do they offer an audience a continuous diet of a particular genre of program. They also become brands, with all the opportunity for extension which that allows. Not many viewers announce as they switch on the television, "I'm going to watch a CBS show tonight," but that is exactly what they say when they tune into channels such as MTV or Nickelodeon. Think of it as the rise of boutiques in competition with department stores.

Boutiques, strongly branded, with low entry costs, can be chains in their own right. The department stores still exist, often by incorporating boutiques, but their market dominance has dwindled.

This situation has a parallel in television. In the United States, where multichannel television has been available for more than two decades, the share of prime-time audiences taken by the networks has declined from more than 90 percent in the late 1970s to around 60 percent at the end of the twentieth century. In homes that subscribe to cable or satellite, the change has been much greater. These families now spend two-thirds of their time with the boutiques rather than with the department stores.

A similar transition is under way in the rest of the world. Once people have more choice, their viewing habits change. In Latin America, for instance, viewing of terrestrial broadcasting in homes receiving cable dropped from more than 80 percent at the start of the 1990s to less than half five years later. Among children, the viewers of the future, the shift to alternative channels has been even more striking.[30] Clearly, loyalty to the networks will erode.

That does not necessarily mean that the networks will disappear – at least not quickly. Indeed, the most remarkable thing about American networks is that they have kept so much of their audience and advertising revenue. In addition, the pace of change will differ enormously from one country to another. It will be slower in Japan and in parts of Europe, such as Germany and Switzerland, that have traditionally had good national television, plenty of "free" choice, and more cultural resistance to American imports than, say, Latin America.

Habit and loyalty will be important. Mass-market television has provided the greatest unifying cultural force since the invention of a common language. Questions such as "Did you watch the game last night?" or "Did you think the president was lying?" form an irreplaceable bond that will help to keep mass-market television alive. Just as a telephone network is more valuable, the more people you can reach on it, so a program is more valued, the more of your friends who want to talk about it the next morning.

Over the next decade, however, people everywhere will begin to think of television not as a choice between half a dozen channels, but as a menu of almost infinite variety. Particular events and programs will still attract immense audiences – larger and more global than today, as access to television continues to spread around the world. But watching television will have become a more individual, personalized, experience.

Paying for rights. More channels mean more competition for viewers. Television companies have long understood the value of an exclusive right to show a Hollywood movie. But in the 1990s live events, and particularly live sporting events, became hot properties. Sports rights are, as Rupert Murdoch puts it, the "battering ram" of subscription and pay-per-view television. Hence, for example, his attempt in 1998, blocked by the British government on competition grounds, to acquire Manchester United, Britain's most successful football club, for $1 billion. Had the sale gone through, it would have been the biggest in sports history.

Having acquired exclusive rights, broadcasters now look for ways to extract from viewers the precise amount they are willing to pay to watch a particular match or show. They will keep a dwindling share of the money they earn, though, because they face a tussle with the athletes or actors who help to make a particular event popular. Sports people will drive hard bargains, squeezing every penny of profit from the television companies; and movie rights will be sold for ever shorter periods and higher prices. The cast of *Friends* demands ever larger fees; footballers want more of the takings from a game. This is because the truly scarce commodity is no longer the channel of distribution but the content that viewers want to see. There, the effect of competition will be to bid up costs for a product whose supply cannot be increased: star players or hit movies.

Changing revenue sources. In the new television world of vast choice and fragmented audiences, viewers pay for content in a wider variety of ways than they once did. In the past, in most countries, either the government or advertisers footed the bill. Now, a rising share of revenues and profits will come from subscriptions.

As broadcast television channels lose audiences, those most affected initially tend to be public broadcasters. Outside the United States and a few other countries, television broadcasting was mainly a public-sector business throughout the first half century of its history. Some public broadcasters, such as Britain's BBC and Japan's NHK, have ownership structures that carefully distance them from government control. All depend mainly on public money in some form or another.

In the 1990s, many of these giants began to suffer the same haemor-rhage of viewers as America's networks, which are their nearest equiv-alents in terms of audience and attitude. In time – although perhaps not until the 2010s – the nature of public broadcasting will have to change. As channels multiply, governments will find it increasingly indefensible to subsidize one or two of them. Moreover, as it will be possible to measure quite accurately who watches what, people who do not watch public broadcasting will demand to opt out of paying for it.

Countries wedded to the idea of public subsidies for television may, instead of financing entire channels, underwrite particular programs considered nationally desirable. Public television will thus become part of the general state budget for subsidizing the arts, rather than a sepa-rate institution.

Commercial broadcast television channels have also been losing viewers. But they have an advantage: they still offer one of the best ways to show advertising to enormous numbers of people. As a result, their advertising revenues have continued to climb. Indeed, the frantic need of dotcoms to establish their brands quickly, and the huge bets their investors allowed them to place, had the ironical effect of allowing old media to profit greatly from new media. Half of the thirty-six companies that advertised during the 1999 Super Bowl, in the most expensive slots ever sold, were dotcoms.[31]

Cable and satellite channels work on a different business model. Their audiences are minuscule by comparison with those of the net-works. Not only do they reach fewer homes; ratings for their programs are almost invariably smaller, and often far smaller, than those for broadcast "free-to-air" programs. In the United States, even the best shows on cable networks reach prime-time audiences of only about one or two million,[32] while the highly regarded CNN has an average hourly audience in the United States of a mere 400,000.[33] In Europe, cable and satellite audiences are often smaller still. As such channels multiply, they tend to cannibalize one another's audiences.

Despite their tiny audience figures, cable and satellite channels are often more profitable than over-the-air broadcasting networks, where these are commercially owned. Not only can they offer advertisers a narrowly defined audience: just children, or sports lovers, or music junkies, say; they also charge their viewers subscriptions. Two sources of revenue are obviously better than one. In the United States, sub-scriptions account for roughly half of all television revenues.[34]

That share will rise with the spread of products such as TiVo and Replay. These are souped-up but simple versions of the video recorder with hard-disk storage, which digitally record programs and have a fast-forward button to enable resistant viewers to skip advertisements. Two-thirds of viewing in homes with TiVo is of recorded programs, and nearly 90 percent of viewers spin through the ads.[35] If viewers refuse to watch advertising, television companies will have to sell more of their product on subscription.

As subscription revenues rise, they will accelerate the transformation of television from a mass medium to a more targeted one. In an ageing society, subscription revenues may bring a particular benefit. Mass-market advertisers want lots of young viewers, who will be big (and adventurous) spenders on consumer goods; thus commercial network channels tend to slant their fare to attract that audience. Older viewers, who make up a rising share of audiences everywhere, may find that subscription channels cater more for their tastes.

The spread of digital television will accelerate these trends. It will be easier to offer many channels, easier to devise clever new features, such as multiple camera angles that viewers can control, and easier to mea-sure what people watch and to charge them accordingly. Television companies may still offer bundles of programs called "channels" – just as music companies continued, when records gave way to CDs, to offer buyers several songs "bundled" on to one product. The question will be whether, if viewers have to pay directly for more of what they watch, they will spend as much time glued to the screen as they did when most programing was paid for by advertisers or government.

Interactivity and the impact of the Internet

Many companies wonder whether television will become the gateway to a new world of interactive services. Of all the electronic communica-tions devices in the home, the television set is by far the most widely owned. Americans own eighty sets for every hundred inhabitants, eas-ily the highest penetration in any large country (Bermuda does better, with more than one set apiece). Around the world, the technology is twice as widespread as the telephone and more than six times as famil-iar as the personal computer.[36] If sheer familiarity were the only thing that mattered, the television set's claim to be the key to the future of communications would be formidable.

The early 1990s accordingly saw hefty investment in interactive tele-vision by telephone companies and some cable-television companies. Many trials were set up, memorably debunked in an article in *Wired* magazine entitled, "People Are Supposed to Pay for this Stuff?" The experiments were killed by a combination of cumbersome technology, high prices, and the arrival of an infinitely more attractive kind of inter-activity in the shape of the Internet. The pay-television services offered by digital broadcasters offer a rough-and-ready sort of interactivity, mainly intended to persuade viewers to spend more on movies, without the enormous investment that true interactive television turned out to require.

Eventually, companies will make that investment, laying broadband cable networks to the viewer's door, and some of the delights once promised by interactivity will become available. The question then will be how people take advantage of a television set that can offer access both to the Internet and to digital television. Occasionally, people may conceivably use the television set as a PC. In Europe, where Teletext is familiar, people are more used to using the television interactively than they are in the United States.

On the whole, though, people will use separate devices for entertain-ment and for straight Internet access. After all, the two kinds of machine have different screens and are used in different ways. The television still sits, in many homes, in a communal room; the PC is more often in a private one. People sit or slouch across the room from a television, using a remote control; they hunch over a computer key-board, clicking on a mouse.

Instead, people will watch video services of several different kinds. They will watch films and other professionally polished entertainment on one kind of device – often elegant, flat-screen, and hung on the wall. Some of that entertainment may be high-definition television, the cin-ema-quality programing of which enthusiasts have talked for so long. It will be sent, perhaps overnight, via the Internet, to be watched by a select audience willing to pay for the best in home entertainment. They might play games interactively on a games console with many other people around the world. (By the start of 2000, more than a million people around the world connected to the Internet via a Sega Dreamcast, a games machine.) And they might use a Webcam – a tiny, cheap, video camera – to make their own on-line entertainment. With Webcams, grandmothers will be able to watch grandchildren, chat-room enthusiasts will be able to smile at the other folk on-line, and – inevitably – amateurs will be able to create home-made porn channels to sell or offer free of charge to a gawking world. All of these many kinds of entertainment will be versions of what was once television. Some, such as high-definition television and games, will make money; others will not.

Another use of interactive television will be in offices, where people already use their screens to watch breaking news, take part in distant training sessions, or monitor conferences from their desks. The broadcasts often run over corporate intranets (a sort of miniature Internet, walled off from the outside world). Ben & Jerry's, a fashionable American ice cream maker, uses Webcasting to inform managers automatically when reserves of Chubby Hubby and other delights fall below designated levels. Large Wall Street brokerage houses regularly use Webcasts starring their own analysts to recommend hot stocks to clients, thus cutting out the television companies' financial channels.[37] In the long run these uses, which pull together aspects of television and the Internet, will probably be the ones that grow fastest.

Radio is already a beneficiary of the Internet. A listing of on-line radio stations in February 2000 came up with a staggering 3,320 radio stations broadcasting on-line, most of them established stations simultaneously offering their free-to-air services on the Internet.[38] Thanks to software such as RealAudio, designed by Progressive Networks in Seattle, listeners can hear sound as it is transmitted, rather than waiting to download it first. As a result, it is possible to run a radio station very inexpensively over the Internet and to reach listeners all over the planet. A transmission can be stored for later listening – creating radio on demand. One early beneficiary has been the BBC's World Service. It has long suffered from poor transmission in many parts of the world. Now its audiences have actually begun to climb, as new listeners have tuned in through the Internet.

For television companies, the Internet is unlikely to be a gold mine. So far it has proved more of a threat than an opportunity, siphoning off audiences, and especially those desirable younger viewers. Most television companies, like movie companies, have built elaborate Web sites. The most successful of these are the sports and news sites, which get lots of hits. Yet none of these uses suggests a convincing way for television to make money from the Internet. Media Web sites are "like building shopping malls in the desert," says James Murdoch, Rupert Murdoch's younger son, who runs the new-media division of News Corp.[39] Time Warner, which was one of the biggest early entertainment investors in the Internet, hopes to use its Web site to develop a mail-order business, rather than relying on advertising to recoup its costs.

The bafflement of the television companies suggests that they may be the wrong businesses to exploit the disruptive technology of the Internet. In January 2000 America Online agreed to acquire Time Warner, thus bringing together the world's largest Internet and content businesses. But new ways to combine the two are more likely to come from elsewhere – from the video-games industry, perhaps, or from sporting-goods firms or travel companies. A rare exception is Endemol, a Dutch television-production firm, which scored a huge success in Britain and the Netherlands in 2000 by combining television and the Internet in the program *Big Brother*. It trained Webcams twenty-four hours a day on a group of people cooped up in a house together, showed edited highlights on television each evening, and allowed viewers to vote (partly on-line) on which one should be evicted each week. For voyeurism, cross-promoted on television and the Internet, there may always be a market. By and large, though, the content that people want on the Internet is simply not the content that the television companies know how to provide.

In a decade's time, television will have become a more diverse medium and will be watched in many more places than it is today. But there will still be television that recognizably descends from the programs people watch today. This familiar programing will be delivered to a distinct sort of machine used mainly for entertainment rather than for shopping or information searches. Indeed, as the quality of screens improves, the television set will become even more different from a PC than it is now. Digital delivery and a flat screen that hangs on the wall will make television feel more like genuine home movies.

Coming home in the evening, the average person will still sprawl on the couch and watch. Viewing choices will increase; some program costs will rise; picture quality will improve vastly; and people will choose from electronic menus. But, well into the first quarter of the century, the main leisure activity will still be essentially a passive pleasure. Couch potatoes we will remain.

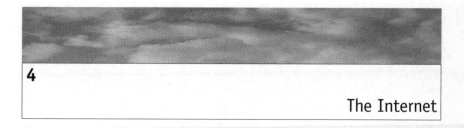

4

The Internet

Never has any new invention shot from obscurity to global fame in quite this way. In 1990, only a few academics had heard of the Internet. Even in 1997, when France's President Jacques Chirac opened his country's new national library and was shown a computer "mouse," he gazed at the curiosity in wonder. Yet by 2000, perhaps 385 million people around the world had acquired a new way to communicate, and a new global source of information on a giant scale.[1]

In addition, the Internet has also created a host of new businesses (some of them evanescent) with exotic names such as Flyswat, Mambo, Egg, Google; an army of new millionaires (some of them temporary); a stock-market boom (and bust); and, above all, the most concentrated burst of innovation the world has ever seen. Many of the astonishing torrent of ideas and business plans unleashed by the Internet will turn out to be junk. But others will transform communications, commerce, and companies. Never in history have so many entrepreneurs attempted, in so short a time, to develop uses for an innovation.

The Internet is thus a global laboratory, allowing individuals as well as the marketing departments of multinationals and academics in top universities to pioneer uses for communications technology. All sorts of experiments, carried out on the Internet, will feed through into other media, changing and developing them. The Internet thus functions as both a prototype and a testing ground for the future of communications. Watching its evolution, we can catch a glimpse of what lies ahead.

The Internet has benefited from the technological revolutions that have slashed the cost of delivering telephone calls and television programs and multiplied the capacity of both types of network. Unlike the telephone and the television, however, the Internet has no established principal use. Instead, it has many uses, including carrying telephone calls and television programs. An open conduit, it is capable of transmitting anything that can be put into digital form.

It offers a peek at the communications future: a world in which transmitting information costs almost nothing, in which distance is irrelevant, and in which any amount of content is instantly accessible – but only a peek, for the Internet is merely a prototype for something more sophisticated. As uses for the Internet multiply, two issues matter. One is how to pay for it and the services it offers. The other is how it will be reached and where the main power to deliver services will reside. The next few years will show the extent to which the answers to these questions can emerge without diminishing the Internet's original exuberance. For the Internet to reach maturity, it must be as accessible to a grandmother as to a geek. It must combine the quality of service of the telephone with the fun of television, at prices that rival those of both. Above all, until people cease to be aware of it as something special and complicated, it will not have found its future.

What Drives the Internet
. . . . .

The early success of the Internet caught many people by surprise. In 1995 it was still possible to write a book on the future of the computer and communications industries and barely mention it. Indeed Bill Gates, boss of Microsoft, did exactly that, devoting only twenty pages or so to the topic in the first edition of his book, *The Road Ahead*. But, by the end of 1996, nearly twenty million Americans a week were logging on to the Internet, four times as many as a year earlier. By the new millennium, the Internet had become an accepted part of many people's working lives, not just in the United States but throughout most of the rich countries of the OECD. [Fig 4-1]

What brought about this transformation?

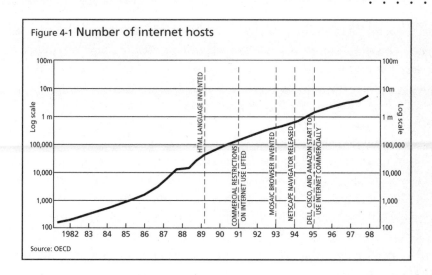

Figure 4-1 **Number of internet hosts**

Source: OECD

An open standard

The single most important reason for the Internet's success is its open standard for transmitting digitized data – voice, video, or text – from one computer to another. The technical protocols that constitute the standard and the software that implements it grew from enthusiastic collaboration among engineers and computer scientists in the public sector and academic research, who constructed a system of extraordinary resilience. More extraordinary still is the fact that, as John Naughton, an Internet historian, puts it, "Not a single line of computer code which underpins the Net is proprietary; and nobody who contributed to its development has ever made a cent from intellectual property rights in it. In the age of the bottom line, where the default assumption is that people are motivated by money rather than altruism, and where software is the hottest intellectual property there is, this is surely extraordinary."[2] There is no gatekeeper, no waiting list, no membership requirement.

Thanks mainly to the single standard, and the browser's role as the standard interface, the Internet is not only transparent and public; it is simple to use. A person learns how to click on the screen – and that is it. No need to retrain to reach the many different services that the Internet can distribute, as proprietary networks require retraining.

· · · · ·

Because the Internet was invented neither by the telephone companies nor by the entertainment industry, no single industry dominates it, forced to defend a pre-revolutionary cost structure or market. It has spread, like every successful revolution, from the grassroots up, driven not by corporate marketing departments but by spontaneous demand from millions of users.

This open quality has been behind the innovative culture of the Internet. In the telephone and television industries, big companies still influence the pace and direction of innovation. The Internet has its giants, too, but it has become a haven for people who, twenty years ago, would have lacked an outlet for their entrepreneurial skills: the cream of young graduates, often with mathematics or science qualifications. This new market has changed the atmosphere in graduate schools. Who wants to admit these days that they do not have a business plan on the go, as well as a thesis? A whole generation of the cleverest young people – people who would once have disappeared straight into investment banks or consultancies – have instead gone through a high-intensity training course in an entirely new kind of business development.

Communication

The Internet is, first and foremost, a way for people to reach other people. Easily its most successful and popular service has been electronic mail, which is almost as old as the Internet itself. Versions of electronic mail existed in the 1960s, but the first e-mail was sent between two machines in 1970 by an engineer named Ray Tomlinson. When he showed his innovation to one of his colleagues, Tomlinson said, "Don't tell anyone! This isn't what we're supposed to be working on."[3] Hunting for a way to separate the name of the user from the machine the user was on, Tomlinson hit upon the @ sign – thus, in the words of Katie Hafner and Matthew Lyon, historians of the Internet, "creating an icon for the wired world."[4]

Electronic mail's wondrous qualities are that it is low tech (no need for fancy software or the latest in modems or PCs); it is asynchronous (you do not need to be logged on when somebody writes to you); it is astonishingly cheap (even in the depths of Africa, keeping in touch by e-mail is less expensive than sending a postcard and vastly cheaper than the telephone[5]); and it is flexible (you can attach an article, a picture, or a song at the click of a mouse). It combines the pleasures of talking and writing. You can reply at once – as you would do to a friend's remark – or you can ponder and write back the following day, as you would doubtless do with a letter. You can write or forward to one or to many, and it is important to think before you click about which you intend to do if you want to avoid embarrassment.

EMarketer, an on-line research group, tried to estimate how many e-mail messages were delivered in the United States in 1998. It came up with a figure of 3.4 trillion. By comparison, 107 billion pieces of first-class mail were delivered in the United States that year. Another study, by the Institute for the Future in California's Silicon Valley, found that the average white-collar worker sent or received an average of thirty e-mails a day.[6] Coupling the ability to send an electronic message with the ease of perfect digital reproduction, electronic mail makes it virtually costless to send many copies to many people. But, although it is costless to the sender, it is far from costless to the recipient. Few people confess to sending junk e-mail; everybody grumbles about receiving it (although one study found that e-mail users typically think they get twice as many messages as they really do).[7] One chief financial officer of a large Silicon Valley firm recalls returning after a week's absence to find two thousand e-mails waiting. In despair, he deleted the lot, unread.[8]

E-mail is not the only way to communicate on-line: instant messaging offers the equivalent of a written telephone call. Users install a piece of software on their PC and enter a list of "on-line buddies" into the program. Once they log on to the Internet, they can exchange instantly displayed messages with any buddies who also happen to be connected. By mid-2000 there were about 130 million users worldwide, sending roughly one billion short messages a day.[9] Most instant messaging goes through America Online, which has fought fiercely to protect the service from outside competition.

The enormous popularity of e-mail and instant messaging suggests that the killer application of the Internet, at least for the moment, is the convenience of a swift exchange of brief notes – and the simple delight that human beings can get from sharing their thoughts with one another. As with so many electronic media, it is communication, not commercial content, that people most crave.

. . . . .

Finding information

Any information that can be stored in digital form – not just data but voice and video too – can be conveyed on the Internet. So the Internet makes it possible for users to become their own librarians, able to research, study, and investigate anything from Marshall's economics to Madonna's music.

By slashing the costs of storing and conveying information, the Internet creates unlimited shelf space, making affordable material that was previously too obscure or expensive to obtain. Some of it is information that has always been in the public domain but was previously inaccessible to most people – because it was held in some special place, or released only to specialists. Press releases, for instance, once landed only on the desks of journalists. Now anybody can read press releases on a company's Web site.

There are also, of course, screens full of dross. "The Net for the first time is causing information overload," says Marc Andreessen of Netscape.[10] Well, not the first time, actually, but it has certainly democratized this curse of the elite. Too much information may be less of a problem than too little, but it is still a problem. Many new Internet businesses aim to help companies that want to shout above the hubbub – or people who want to filter out unwanted noise.

Because access to the Internet is open and virtually free, it offers vastly more information and content than any other electronic medium. As a result, people desperately need some way to organize the stuff so that it can become useful. That is what search engines and directories do. Their crucial importance explains why Yahoo!, a directory, is the Internet's most visited site. Indeed, so many search engines now compete to help that there are guides on how to search for the right search engine, one of which, Direct Hit, ranks its suggestions on the basis of other searchers' behavior.[11] Armed with a good search engine, it is possible to collate information that once could not have been economically brought together, or to seek out niches, whether for people or products.

But search engines are not yet good at distinguishing between, say, a press release and a detailed news report. They find it difficult to know which page will most interest a particular user. New kinds of software are being developed to solve such problems. By learning a user's interests and tastes, search engines will be able to target the most appropri-ate sorts of information. They will also know more about language, so that users can put their questions in a more natural way.

Security

The main reason for the Internet's success and charm is that it is an open network – or rather, a network of networks – across which any-body can send any information. But this openness creates headaches for those who want to use the Internet for valuable transactions. The main threats are from hackers and viruses, which can both use the Internet as an entry point into a company's computer network and then intercept in transit material such as credit-card numbers and pri-vate information. But, even without such intrusions, commercial users need privacy and security, two qualities that are difficult to incorporate into an open network. Encryption and authentication offer answers, but encryption alarms law enforcers and authentication will not work better until there is a common international agreement on which variants to adopt.

Hackers and viruses

The threat from hackers and the costs of keeping them at bay are both growing rapidly. First, thousands of proprietary systems that would once have been understood only by a handful of experts within a com-pany are being replaced by the universal currency of the Internet proto-col, resulting in vast returns for the hacker. Second, individual employees are acquiring more power to sabotage, because sensitive information about customers can be copied without trace and electron-ically distributed. Third, the sheer amount of damage a hacker can do has risen as industries have become more and more dependent on their computer systems. With larger systems comes ever greater potential for mutilating taxation and social-security records, disrupting air-traffic control, interrupting electricity supplies, or stealing money from banks.

Nobody knows how much hacking takes place. But most attacks – 95 percent, according to America's Federal Bureau of Investigation – go undetected, and of those that are discovered, only about 15 percent are reported. Microsoft itself sheepishly reported, in the autumn of 2000, that it had been infiltrated by hackers. In fact the truly remarkable thing is how little alarming damage hackers seem to do. This might be in part down to the hefty sums companies spent on tightening security and overhauling their networks in the months leading up to the end of the millennium. Although failures caused by the clock rolling over to a year ending with the digits 00 were rare, the investment may have left some lasting benefits.[12]

Computer viruses – rogue programs that can disable a computer or destroy data – also seem to be a growing problem, if only because there are ever more PCs for them to infect. Some viruses are planted deliberately and maliciously by a hacker; others are introduced unwittingly by a user. The ubiquity of Microsoft software also widens the opportunities for viruses to spread; this was clear in May 2000 when an e-mail message declaring "I love you" and carrying a malicious attachment stormed around the world, clogging servers and destroying files. The so-called "love bug" installed itself on a hard disk and e-mailed itself to all the addresses on the Microsoft Outlook address book. Non-Microsoft software – such as the Macintosh and Linux operating systems and Lotus Notes – was immune from infection. The moral: monoculture, in computer networks as in agriculture, increases the risk of infection. Biodiversity is safer.

Interception and encryption

Encryption offers a way to protect material travelling across the Internet, whether a private message, a purchase order, or a credit-card number. Encrypted information – information which is scrambled into a set of numbers that can be decoded only by someone with an algorithm or key to the code – can travel across an open channel or be stored on a hard disk, concealed from those without the necessary key. Encryption thus gives protection both in transit and in storage. It offers an answer to some of the trickiest problems created by the Internet, such as the desire for privacy. But it also aggravates other tricky problems, such as the ease with which the Internet can be used by illegal groups.

The trouble is, the requirements of electronic commerce tend to clash with those of domestic security. Encryption robust enough to prevent hackers from stealing a software program could also be used by criminals or spies. Many governments fear that the Internet makes crime and spying easier, and therefore want to snoop on their citizens on-line. Britain, for example, built in 2000 a new e-mail surveillance center with the power to monitor all e-mails sent to or from Britain. Internet service providers (ISPs) will be obliged to install cables that will download material to the secret service.

Because encryption makes such intervention harder, governments have tried to restrict it. The United States government initially tried to limit the export of "strong" encryption technologies that are particularly difficult to decode. A growing number of other countries go further and limit the domestic use of encryption. The United States, however, largely abandoned its policy in early 2000, by which time several companies in foreign countries had developed strong encryption products to fill the gap left by the American ban.

Apart from the problem of securing on-line messages against snoopers (government or private), transactions on the Internet also often need authentication: some way to prove that they have indeed been sent by the person who claims to be their originator. A number of countries have passed laws to recognize one authentication technique, that of digital signatures. But authentication used in one country is often not recognized in another. If the Internet's full cross-border potential is to be achieved, that will need to change.

Governing the Internet
. . . . .

The Internet's early apostles liked to present it as a sort of open frontier where no writ ran. "The Net interprets censorship as damage and routes around it," John Gilmore, an on-line activist, once said. But the Internet itself follows clear conventions. Like every other way of communicating, it has had to evolve common principles to work smoothly.

Moreover, national standards are not enough. Like the telephone or airlines, the Internet needs global rules.

Despite the Internet's lack of a central source of authority, there are bodies that run it, performing two main tasks: developing technical standards and administering the system of domain names. These bodies are an odd lot, self-governing and run mainly by enthusiastic volunteers, some of whom have been around the world of the Internet since its birth. They rarely have a fixed membership: "no cards, no dues, no secret handshakes," boasts a notice on the Web site of the Internet Engineering Task Force (IETF), the main standard-developing body. They meet a few times each year (face-to-face: even the pioneers of the on-line world occasionally need to shake hands), and they tend to take decisions largely on the basis of consensus.

These self-appointed oligarchies have worked well through the early decades of the Internet's evolution because they are made up of like-minded individuals – largely American, mostly scientists and engineers – with common interests and a common approach to solving common practical problems. Once they run up against issues of economic and commercial policy, however, the consensual process tends to falter. In 1999 the IETF was riven by a bitter dispute over whether or not the organization should help law enforcers to conduct wire-taps. Some members, reflecting the anarchic, anti-authoritarian approach that has always been one of the charms of Internet pioneers, wanted no truck with Big Brother. Others, representing companies that make telecoms equipment, feared that their products might be required to comply with Federal wire-tapping laws.

The dispute reflected two fault lines in the governance of the Internet. The first is between the core Internet community and the people from the world of telecommunications, who come from an altogether more top-down, centralist, big-company culture. Although the two worlds have to some extent converged, they still sometimes pull in different directions. The second division is between the United States and the rest of the world. American dominance of the Internet means that other countries often see its rules as having been set mainly in America's interests.

A second standard-setting body deals with the main application that runs on the Internet, the World Wide Web. Founded in 1994 by Tim Berners-Lee, the Englishman who invented the Web, the World Wide

Web consortium, or W3C, is an international industry consortium which over the years has developed twenty technical specifications for the Web.[13] Like the IETF, though, it increasingly deals with questions of public policy. It developed, for example, an Internet rating system known as Platform for Internet Content, or PICS, intended to help parents and schools filter material unsuitable for children. That embroiled it in a controversy with civil libertarians, who worry that such filters can equally well be used by disagreeable governments to restrict their citizens' access to the Internet.

A third body, the Internet Corporation for Assigned Names and Numbers (ICANN), was born in controversy and has been up to its eyes in it ever since. Its main task is to coordinate the Internet's highly centralized system of numeric addresses: only thirteen giant "root servers" know where one computer has to go to find the address of another. That gives it vast potential power. In addition, it manages the domain-name system, which turns those numbers (207.87.8.50, for example) into human-friendly monikers such as *www.economist.com*. This is the trickiest of all the tasks that governing the Internet entails. Because only one user in the world can claim any given domain name, the whole subject has been a minefield of trademark disputes (see Chapter 9).

Unlike the IETF, ICANN did not grow organically from the grassroots up. And unlike W3C, it was not the initiative of a single Web pioneer. Instead, it was set up in 1998 to end a row over the way the allocation of electronic addresses was managed. For many years that job was done by an American company, Network Solutions Inc (NSI). NSI handled the registration of domain names as a monopoly and began, in 1995, to charge for the service, which had previously been free. That led to grumbles, which grew louder with the trademark disputes that arose from its first-come, first-served allocation policy. To resolve the situation, Ira Magaziner, President Clinton's adviser on the Internet, came up with a plan that replaced NSI with ICANN, an organization that itself developed out of an earlier Internet body run from Los Angeles by Jon Postel, a pioneer of the Internet's forerunner, the ARPANET.

ICANN has had one considerable triumph: it has converted the registration of domain names from a monopoly into a competitive business, with more than 120 registrars now accredited to sell them. But it has an unwieldy, nineteen-member board; a plethora of working parties and "supporting organizations;" no stable source of funds; and an uneasy relationship with government. As a hybrid – part on-line community and part real-world international organization – it is a compromise between the old, free-wheeling world of the early Internet and the new commercial world, dominated by money and the law. Ultimately, the question is whether the remarkable arrangement, whereby the most important areas of Internet governance are handled by private (and mainly American) bodies, can survive, or whether governments or an international organization eventually move in.

Access to the Internet

Two linked questions hang over the Internet's immediate future. How will individual users reach it – how long, for instance, will the PC remain the principal point of entry? And how will they pay for their access? At present, the experience of users differs enormously, depending on where they live.

Access: connecting the home

All over the world, the telephone line's pair of twisted copper wires is the usual connection from home or small business to the Internet. But it was designed many years ago to carry two human voices – not streams of voice, video, and data – and it transmits relatively slowly. Dialing up a local ISP each time you want to check your e-mail or look at a Web site is painfully slow.

Telephone companies and cable companies hope to change that by providing broadband, high-capacity access good enough to carry video services. Telephone companies are using a technology called Asymmetrical Digital Subscriber Line (ADSL) to increase the carrying capacity of copper wire and enable it to carry video services, or upgrading connections with fiber-optic cable coupled with digital switching and transmission. So far, though, the exercise has been slow in most parts of the world. Only where many people live in apartment blocks, as in Hong Kong, is it possible to connect them relatively inexpensively. In the

United States, in 2000, only 2 percent of homes had broadband links.[14] Even by 2005, the proportion is unlikely to be more than 20 percent, and in Europe it will be even lower.

The problem is largely cost. To upgrade a telephone system to a speed at which it can carry adequate video costs around $1,000 a home. Companies are making such investments only when they can see an average return per home passed – not just per home connected – of at least $200 a year. If only half the homes passed are likely to buy the service, then the builder needs to be sure of making $400 a year from each subscriber.

Such figures are not unimaginable. After all, the average American home spends roughly that on a cable-television subscription and more than twice as much on the telephone each year. But both of these are well established services. The Internet might require many years for revenues to build up to such levels. Telephone companies, which tend to charge about $40 a month for ADSL, usually find that fewer than one in five households in a neighborhood want it. Yet, for the infrastructure investment to pay off, investors must be sure that people (or advertisers) will pay such sums over and above what they already pay for television or telephone calls. Small businesses may do so quite soon; homes will take much longer.

Where customers do have high-speed access, the early evidence suggests that they make more use of the Internet. Neilson/Netratings, an on-line survey, reported in January 2000 that users with high-speed connections visit the Internet 83 percent more and view 130 percent more pages than those with slower access. Of course, the explanation may be simply that the most enthusiastic Internet users are the ones who splash out on lots of bandwidth. But another study, carried out in Europe by Chello, a company offering high-speed access in six countries, found that subscribers to its networks spent seventy-two hours a month on-line, twice as much time as dial-up users. It also found that they went on-line more frequently and – important for the future of e-commerce and the financing of such networks – that they had spent $457 on-line in the previous six months, compared with $275 for dial-up users over the same period.[15]

The crucial selling point of broadband access may turn out to be not that it delivers good-quality video, but that it can be always on. But new Internet-enabled mobile telephones will have that property too.

Wireless systems are always less expensive to install than those requiring lots of wires or cables. Companies thinking of investing heavily in broadband networks for homes and small businesses might pause to see first what people do with their smart new cellphones.

Access: the PC or...

Almost everyone who uses the Internet, whether in the office or at home, does so on a PC (or maybe an Apple Mac). Wonderful though it may seem to have the power of the Apollo 13's main processor on your desktop, the PC is complicated to use, compared with the elegant simplicity of a telephone or television set. The industry sometimes boasts that, had cars advanced as quickly as the processing power and memory of the PC in the quarter century of its existence, they would travel at supersonic speed and cost only a few dollars. In fact, the PC is all too like that automobile of 1910: it constantly and inexplicably breaks down and requires much time and skill to keep it on the road.

The sheer versatility of the PC will ensure that people who do lots of work from home will still have one to design their PowerPoint presentations or manage their finances. However, most people will want something more modest, either in addition to, or instead of, a PC. They will choose from lots of specialist gadgets designed for particular purposes, such as television set-top boxes, games consoles, telephones with screens, various kinds of handheld computers and personal digital assistants, and smart mobile telephones. Many of these gizmos will have a wireless, rather than a fixed, connection to the Internet. They will be designed to do fewer things than PCs, but to do them well.

IBM calls this the dawn of "pervasive computing" – in other words, computing can take place anywhere, any time.[16]

At least some of the software that these new devices use will sit, not in their memories, but on the Internet itself. The Internet will thus acquire another function: it will become, in the jargon, a computing platform. Many software services are already available on-line, such as e-mail (Microsoft's hotmail is one example), and several firms offer word-processing or spread-sheet applications. In future, computers themselves will have direct access to Web services, in effect turning the Internet into a giant computer. Just how many computer applications will move from the PC to the Internet is the subject of hot debate in the computing industry. Probably it will vary from one device to another.

For companies, as for individuals, this new era will make life simpler. PCs absorb huge amounts of corporate time to maintain, debug, and monitor. They allow users to do things in the office that employers would rather they did at home, from downloading music to looking at naughty pictures. The "one-size-fits-all" technology of the PC may then give way to more specialized and diversified appliances, all connected to a common network. If so, help desks and systems managers will heave a sigh of relief, and so will many users.

When things start to think

Some of the objects connected to the Internet will not be deliberately operated by human beings at all. They will be "thinking things," in the phrase of Neil Gershenfeld, a senior academic at MIT's famous Media Laboratory.[17]

Combined with wireless, the Internet can feed into parking meters, drinks dispensers, ticket machines, and other gadgets in streets and offices. It can monitor machinery in the home (is the security alarm working? Does the cooker need a maintenance call?) or in offices.

A whole host of science-fiction gizmos is under development: Panasonic, a Japanese consumer-goods company, has built a microwave that can respond to cooking instructions delivered on the Internet. Microsoft has a refrigerator that keeps track of groceries and makes its own shopping list.[18] Samsung of South Korea is developing a microwave that can read the barcode on the side of pre-packed food and adjust its cooking program automatically.[19] And Electrolux, a Swedish company which introduced the modern refrigerator and still sells more of them around the world than any other company, has created the ScreenFridge, with a flat computer monitor on the door. The monitor can be linked to a camera to watch a sleeping baby or check who is ringing the doorbell; or, say its designers hopefully, it might run advertisements to lower its cost to consumers.[20]

Such devices may initially have a limited market. People are likely to be frightened off both by the chance that some vital appliance will break down (imagine the horror of a central-heating boiler that was as prone to crashing as a PC) and by the implications for security and privacy. But that fear may eventually be overcome if a truly useful gadget emerges. From the point of view of appliance manufacturers, such creations present new opportunities to retain a lasting link with the customer. The maker can offer a lifetime service, monitoring whether the device is using power at the cheapest time of day or whether the settings are correct, and can also make a new sale by noticing when the device is approaching the end of its life. Electrolux is taking that idea to its logical conclusion and experimenting with a pay-per-use washing machine. People pay $55 to have it installed and about $1.12 a wash, a fee that appears on their monthly electricity bill. One aim is to encourage energy efficiency. At the end of its life, the machine is replaced and recycled.

Plenty of pieces of machinery may eventually work better because they are linked to the Internet. But, if things think, people will want them to do so reliably, in useful, money-saving ways – and to keep their thoughts to themselves.

Paying for the Internet

One of the striking aspects of the Internet is that countries with similar levels of education and wealth use it with different degrees of enthusiasm. Among the rich OECD countries, a "digital divide" has opened between countries where English is the first language or widely spoken as a second language, and the rest – mainly central and southern European countries and Japan.

What could explain the difference? Perhaps it is culture. Americans tend to drive straight home after work rather than gossiping in bars as many Europeans do. They have large homes with separate dens, ideal for sitting and surfing on a PC, whereas Europeans with their smaller homes have to clear the clutter off the kitchen table and plug in the computer before they can do the same. Or perhaps it is climate. Scandinavians need indoor activity on dark winter nights, whereas Spaniards and Italians can still go for a stroll and an aperitif.

Finding the explanation has become more important for European countries and for Japan, as they have become more aware of the extent to which their Internet development lags behind that of the United States. That gap seems to be growing wider, not narrower. Research by the OECD reveals that the United States is pulling away from the crowd, particularly according to those indicators that measure growth in electronic commerce. Between September 1999 and March 2000, the United States added an additional 25.1 Internet hosts for every thousand inhabitants. (The number of "host" computers, the powerful central computers on which Web sites and other information are stored, are the most reliable measure of the scale of the Internet in any given country.) By contrast, Britain added 5.5 hosts, Japan 4.1, Germany three, and France 2.7.

What is happening? The answer seems to have little to do with culture or climate and almost everything to do with the way people pay for Internet access.

The impact of all-you-can-eat

The crucial evidence that pricing matters comes from a study, again by the OECD, of two connected indicators – the prevalence of radio stations on the Web and the amount of multimedia content, such as music and short films – relative to population. The United States, together with Canada, Australia, and New Zealand, are in a completely different league from other rich countries on these two counts.

The main difference between these four countries and the rest of the OECD is not that their overall costs per hour on-line are lower; in some cases, they are higher. It is that they have unmetered access to the Internet. At the time of the OECD's study, domestic and small-business Internet users throughout Europe paid the telephone company for every extra minute they stayed on-line. In the United States, Canada, Australia, and New Zealand, the Internet truly feels "free" because, once on-line, the clock does not tick relentlessly away. Although, by paying a lump sum for unlimited telephone time, they might end up actually paying more than they would in some ticking-clock countries, people in these four countries clearly prefer things that way. They feel more comfortable about staying on-line to hunt for information, take part in auctions, or monitor stock prices.

The effect on attitudes is especially visible in the development of Internet radio stations. The idea of using the Internet to listen to the radio for more than a few minutes would seem madness to most Europeans: why run up a telephone bill listening to something that might be freely available over the air? In the United States, by contrast, on-line radio is booming. A survey in July 1998 found that 6 percent of American users had tuned to Internet radio. Six months later, the proportion had shot to 13 percent. And the leading radio stations found that average on-line times were running at seven hours a month as people kept "streamies" (streaming audio) running while they did other tasks on-line.[21] [Fig 4-2]

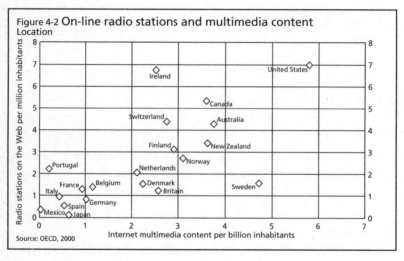

Figure 4-2 On-line radio stations and multimedia content

Source: OECD, 2000

Whenever Internet users switch from metered to all-you-can-eat access, the amount of time they spend on-line jumps. The best known example came in December 1996 when America Online introduced flat-rate pricing. Initially, it was chaos. In the first few months, AOL was overwhelmed as people began to treat it like television: switch on and stay on. The worse the congestion, the greater the incentive for people who had got on-line to stay on, making the jams even worse. Within a year, though, the service had improved. Time spent on-line by the average customer zoomed from fourteen minutes a day in the third quarter of 1996 to forty minutes a day a year later, and to sixty-four minutes a day in the first quarter of 2000.[22] Something intriguingly similar occurred in New Zealand. Xtra, Telecom New Zealand's ISP, introduced flat-rate pricing in May 1999 and the average time users spent on-line doubled, from nine hours a month to eighteen, by the end of the year.

In Britain, early in 1999, Freeserve, an ISP, began to offer free Internet access. Several other European ISPs soon copied. The service has been popular, but it has made its money by taking a slice of the telephone tariff that users pay. Some new entrants to Britain's telecoms market, such as two cable-communications companies, NTL and Telewest, then went further and introduced flat-rate, all-you-can-surf Internet charges, including telephone costs, for a flat monthly fee. Telewest's product was based on research that found that 32 percent of people in Britain who had not yet used the Internet would do so if offered a flat rate. In just over two weeks, Telewest's service picked up more than 100,000 subscribers.

Such pricing schemes go against the grain with many telephone companies, which argue that their networks were designed for short chats, not long surfs. In the United States, where telephone companies have benefited enormously from the growth in demand for second and third telephone lines for Internet access, they have also suffered from the congestion that calls create as they clog networks. Initially, American and Australian telephone companies greeted the Internet's popularity by demanding the introduction of metered rates, to keep surfers at bay. Much was written about the "tragedy of the commons," a metaphor that describes the way free access to a common resource results in overuse.

In fact, illogical as it may sound, people might be willing to pay a premium for a flat rate. In the United States, where residential users often have a choice, many folk opt to pay more under a flat-rate plan than they would do for a plan based on use. This is not new: in the 1970s, when the Bell System experimented with various ways to pay a telephone bill, even people who hardly used their telephone generally opted to stay with a flat-rate plan.[23] The reasons, it appeared, were partly a desire to avoid the "mental transaction costs," as one economist calls them, of worrying whether it was worth talking for an extra few minutes, and partly a widespread tendency for people to overestimate how much they used the telephone. Most of all, though, people want the peace of mind of knowing that they are protected against large bills. In other words, they are willing to pay extra to avoid paying by the minute.[24]

The implications for mobile

The mobile telephone will become, for many people, a gateway to the Internet in the first decade of the century. The development of wireless applications protocol, or WAP, has created a global standard for adapting Web-based services to cellphones. In the first months of 2000, WAP phones were more hyped than bought, and those who did buy found them dismally slow, expensive, and confusing to use. Like dial-up access on an ordinary telephone, it requires the user to connect each time. The huge success of NTT DoCoMo's i-mode service springs partly from the fact that it is always on. Its bandwidth may be tiny compared with that of a WAP phone, but the service runs non-stop, like an office Internet connection.

That will be cured with the arrival of "third-generation" (or 3G, in the jargon) mobile services, which will have more lavish bandwidth. Before these are available, users in Europe will have an intermediate option: so-called 2.5G, based on a modest software upgrading of the base stations of Europe's GSM operators. Many people may find this a perfectly adequate way to combine the Internet with the mobile telephone. They will use such telephones differently from a PC – to look at time-sensitive information such as share prices or betting odds, perhaps, or to carry out simple transactions such as buying theater tickets. They may also simply talk for longer. At present, people in the United States chat on their cellphones for about six or seven minutes a day – compared with an hour or so on their wired telephones. Indeed the mobile operators may find to their surprise that old-fashioned chat brings in more revenue than fancy new information services.

How fast the new telephones catch on, however, and how much they are used, will depend on pricing. Mobile telephone companies, even in the United States, generally bill users by the minute. Yet users of mobile Internet services are likely to be just as sensitive to the structure of charges as those who stick to their PCs and landline.

The pressure for change will come first in North America and the other flat-rate countries. In May 2000 AT&T announced a pricing package offering users of mobile Internet services unlimited access for a flat monthly fee of $14.99. Significantly, North American operators also tend to charge flat rates for unlimited use of messaging systems. In Europe, by contrast, the convention is to charge per message. This is partly because operators worry that an all-you-can-eat pricing structure would gobble up some of their lucrative market for voice calls. Contrariwise, a move to flat-rate pricing for fixed lines will add to the pressure for a similar pricing structure for mobile services.

Some European politicians hope that Europe can use mobile communications to leapfrog ahead of the United States in Internet access. And not just politicians. In 1999 Larry Ellison, chairman of Oracle, told a conference (in Europe, of course) that the WAP telephone would allow "more Europeans to have access to the Internet than Americans."[25] However, the lesson of pricing for fixed-line access suggests that this will be possible only if European mobile-telephone companies are willing to offer all-you-can-eat access right from the start.

The dilemma for the telephone companies, in short, is how to exploit two different markets over a single network. The best growth prospects are with the Internet and the myriad services it delivers. But, for the moment, the big money is in voice calls. As so often with the arrival of the Internet, the question is how fast to cannibalize an existing business in the hope of building a new one on top of its carcass.

Why the Internet Matters
. . . . .

The Internet is, in one sense, merely an enormously efficient way to transport digital data around the world. In another, it is a laboratory for communications in the future. It has, however, three particularly important characteristics.

A global span

"There has never been a commercial technology like this in the history of the world, whereby from the minute you adopt it, it forces you to think and act globally," says Robert Hormats, deputy chairman of Goldman Sachs International.[26] This extraordinary quality will grow more important as the rest of the world starts to catch up with the United States. At the start of the new century, more than half the Internet's users were already outside the United States, but America still had by far the largest number of users as a share of regional population. This imbalance between America and the rest of the world means that the Internet's potential role as a global medium was barely apparent. The telephone still connects continents more effectively. But that will change. Thanks mainly to the convergence of the Internet and the mobile telephone, use outside the United States will soar in the coming decade. [Fig 4-3]

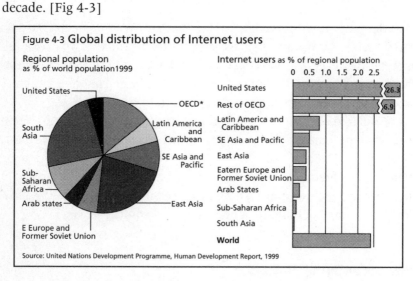

Figure 4-3 **Global distribution of Internet users**

Source: United Nations Development Programme, Human Development Report, 1999

Among the winners, developing countries will be especially important, for they will enjoy new freedoms: a way around overpriced international telephone and postal services, for instance, and a short-cut to information that may not be available locally, such as scientific articles and uncensored local news.

Once it has true global reach, the Internet may become the main platform for international contact. It provides a shop window in which a company can display its wares to a world market. It offers a chance for people from different countries to swap information and ideas. It provides the means for people who are cut off from the world by censors and oppressive governments to tell their stories. No other innovation has ever had quite such earth-shrinking potential.

Convergence

The Internet will change electronic products of all kinds, from television and the telephone to games and cameras. In the case of the telephone, its most dramatic impact will be on rates. "I'm not sure what the Internet is good for," said Bill Gates during its earlier days, "but I don't know why you would want to be in the long-distance market with that thing out there."[27] A number of companies offer telephone calls over the Internet and, while the sound quality is often no better than could had from tying two cans together with a piece of string, the prices are low enough to drag down long-distance charges. In addition, the Internet provides a way to try out new kinds of service, such as the videophone (long predicted, finally affordable, thanks to the Webcam).

In the case of television, the Internet offers a bridge between the continuous flow of information (the television model) and the focused search for facts (the Web site model). The Internet will drive uses of television and radio in the work place, something that big broadcasters have never cared about.

Above all, the Internet will be integrated into other products. It will be part of the telephone service, part of the way a television works, part of a games console. It will connect things and people and animals and companies. People will stop thinking of the Internet as a separate entity and be aware only of the services it delivers, not of the network itself.

Innovation

Above all, the Internet has become the most powerful driver of innovation the world has seen. Because of its open, flexible protocol, thousands of companies, founded by the best-educated bunch of entrepreneurs ever to blitz a business, are making (and, periodically, losing) huge sums of money developing new ways to use it.

One result has been to change the structure of the communications industry, shifting the focus of innovation away from the old giants and toward these young hothouses. Another has been to drive forward communications technology at a formidable pace. Through the Internet, new products can be developed and launched relatively inexpensively, potential customers and investors can be targeted, and markets can be quickly identified and tested. No other mechanism provides such instant links between inventors and their customers.

The host of Internet start-ups that are sprinkled around Silicon Valley, through San Francisco's North Beach, in Munich's periphery in southern Germany, and in the Cambridge heartland in the east of England, form a vast R&D laboratory. They bet their future and their employees' share options on a single bright idea. Driven by ample supplies of venture capital and staffed by bright graduates or ex-employees of larger firms, a few will come up with ideas that work. Often these will be snapped up by bigger high-tech companies, such as Microsoft, Intel, and AOL, which can roll out products swiftly to a mass market.

That, above all, is why the Internet matters. It makes the market system – since the collapse of communism the dominant method of allocating resources around the world – work better. Those parts of the world that embrace the Internet will find themselves better able to compete than those that lag behind.

Consumers and Electronic Commerce

When, in 1888, the Sears and Roebuck Company published its first mail-order catalogue, it was an immediate success. Over the next five years, catalogue shopping bounded up by at least 25 percent a year. Hundreds of local hardware and clothing stores went out of business – "disintermediated," as today's jargon would put it, by a company that could create more direct links between wholesaler and customer. But once the first flush of enthusiasm for this astonishing new way of buying wore off, the growth of mail-order catalogue shopping settled down to about a tenth of retail sales in the United States, and less than that in other countries.[1]

Will that be the future of Internet commerce too? Many think it might be. Sceptics notice, indeed, that some on-line retailers have actually begun to emulate Sears Roebuck and produce paper catalogues. Unlike a Web site, catalogues sit around the house, reminding people to send in their orders and catching visitors' attention. Sceptics also note that several on-line retailers are opening bricks-and-mortar stores, and that the level of on-line sales is diminutive. Even in America, home to almost all the e-commerce that matters, it accounted for less than 1 percent of retail sales in early 2000. By 2004, after a decade of lavishly financed growth, global electronic commerce between businesses and consumers will probably still be worth no more than $450 billion, or less than three times Wal-Mart's total sales in 1999.[2] And that assumes no recurrence of the debacle of Christmas 1999, when many on-line retailers spent heavily to attract customers and cut prices, and then made a hash of fulfilling orders. If this is the future, the sceptics might say, bring back mail order. [Fig 5-1]

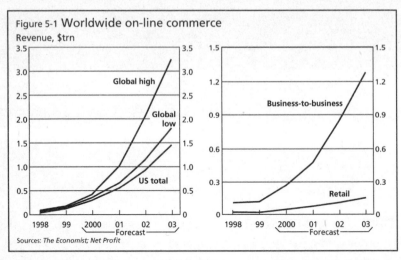

Figure 5-1 **Worldwide on-line commerce**

Revenue, $trn

Sources: The Economist; Net Profit

In fact, although the Internet will have a more dramatic effect on commerce between businesses, and on corporate structure (the subject of the next chapter), it also has enormous potential to transform commerce between consumers and businesses. It does two main things. It allows prices to be set in new ways. And it helps buyers and sellers of products and services to find one another, increasing choice. The second may turn out to be even more attractive to shoppers than the first.[3]

The Internet thus gives consumers new power: to compare, to choose, to bargain, to demand exactly the products they want. In businesses as diverse as stock trading and car retailing, banking and bookselling, tourism and gambling, the communications revolution is introducing new competition, holding down prices, expanding choice, and boosting brands. It will influence bricks-and-mortar companies, changing the ways they do business. It will also help companies to move beyond mass manufacture to produce goods and services shaped to meet individual customers' requirements, a process dubbed "mass customization." It will cut the costs of serving vast markets, opening the way for a stream of niche products. And, eventually, it will alter fundamentally the structure of many consumer businesses, eliminating some and creating others.

So the Internet offers more than a new way to sell directly to customers, for which the telephone will long remain more important, or a new advertising market, a task that television will continue to do better. Rather, it is a distribution and payment channel. It offers information (on prices, choice, and consumers' behavior); it is a marketplace, experimenting with entirely new ways to trade; and it is a way for even the smallest business to acquire global reach.

From Clicks to Bricks

Using the Internet to sell things to consumers is an idea which, by the start of the century, was only about six years old. But, as the idea spread, it became the Internet's driving force. Just as toothpaste advertisements once persuaded people to brush their teeth, so the flurry of advertising for dotcoms and their products has drawn the Internet to the attention of people who might never have taken an interest in it.

This sudden novelty had much the same effect as Dutch tulips had in the seventeenth century. It sparked the boom in dotcom shares in 1998-99. At its peak, in 1999, it produced the valuations of a classic bubble. Amazon.com's capital value overtook that of all America's off-line bookstores, including Barnes & Noble and Borders, put together. America Online, America's leading Internet service provider (ISP), acquired a market value greater than that of General Motors and bought up Time Warner, previously the world's largest media empire. Yahoo!, the search engine, was more valuable than Boeing and traded at more than 150 times the value of its sales. These valuations were (loosely) based on the assumption that the consumer Internet and the new companies that populate it would seize markets and profits from old-economy companies.

In fact, as the bubble subsequently popped, it became apparent that e-commerce would change the landscape in a different way. Some of those new companies will indeed seize markets. But doing so often requires a new brand to be established. And although the Internet has created a cornucopia of brands, getting customers to notice them has proved harder. Open a high-street store, and the shop to some extent advertises itself. In the jumble of the Internet, it is harder to stand out. In the run-up to Christmas 1999, which many business-to-consumer (B2C) dotcoms regarded as make-or-break time, some newcomers spent three-quarters of the cash they had raised from venture capitalists on marketing.[4] Equity finance, which conventional businesses would have invested in bricks, was splurged by dotcoms on trying to buy clicks.

Although many dotcoms have burned their way through their investors' millions at a hair-raising pace – before it shut down in summer 2000, Clickmango, a British on-line health care business, was making sales of £2,000 ($3,000) a week, but only by spending £25,000 of its investors' cash a week[5] – plenty of old-economy companies have been hunting for ways to incorporate the Internet into their existing activities. That alone will drive electronic commerce from exotic to mainstream in the first decade of the century.

Traditional retailers find it difficult to add the Internet to their other ways of reaching customers. The integration disrupts established culture and eats into the revenues from other channels. But the companies that dominate the Internet will increasingly be conventional businesses that find imaginative ways to become true "multichannel" retailers. As on-line retailers are also learning, it is usually better to have bricks as well as clicks.

Potential winners

Certain businesses have taken to the Internet like ducks to water. Some of those are obvious; others less so. It was clear early on, for example, that the Internet and pornography were made for each other. Not only can the Internet offer instant gratification in a way that old-economy shopping cannot; it provides privacy, anonymity, and breathtaking choice. No wonder porn quickly became one of the most valuable on-line commercial activities.[6]

Less obvious was book-selling. No instant gratification here: buy a book in the shop around the corner and you are halfway through the first chapter before Amazon.com has passed your order to the courier company. Yet books and music have become the most popular on-line shopping items. The reason is partly that many bookstores cannot offer instant gratification either, unless you are buying one of a few best-sellers. It is simply too costly to carry thousands of individual items in stock. In addition, books are commodities (although authors hate the thought). Wherever you buy *The Death of Distance*, you will get identical words, arranged in the same order.

Most of the products that have done best on-line have the character-istics of porn or books or both. Thus software, like naughty pictures, can be inexpensively marketed and distributed on-line. The same applies to airline reservations, which rival software as the most popular on-line purchase, and stockbroking. Both airline reservations and shares have long been bought over the telephone, and the Internet adds instant delivery.

Initially, the products that sold best on-line were those that appealed to the affluent, well educated young who made up the bulk of early Internet users. Thus in 1999, according to a survey by Shop.org and the Boston Consulting Group, nearly 18 percent of the computer hardware and software sold in the United States, as well as almost 15 percent of financial brokerage services and almost 9 percent of books, had already shifted on-line.[7]

As the Internet attracts a broader social range, so commerce will widen. The whole business of shopping electronically for some prod-ucts, such as movie tickets, will grow more attractive as mobile tele-phones encourage on-line ordering. Increasingly, baffled customers can click on a button on a Web page to get human help from a call center. Such combinations of telephone and Internet will give a fillip to on-line banking.

New technologies will make it more attractive to buy "high-touch" products on-line. Digital distribution will improve (the Rocket e-book already allows a shopper to download a new book at the click of a but-ton). Physical distribution will get better too, partly because Internet retailers are starting to build their own stores and depots.

Above all, conventional retailers will put their muscle behind elec-tronic commerce. As they do so, though, they will find threats and problems at every turn.

Possible losers

Where a product can not only be sold, but be easily distributed on-line, electronic commerce will pose the biggest threat to traditional retailers. But it is not the only threat. Many established businesses have some activities that might be picked off by on-line rivals. Newspapers, for example, may continue to exist in physical form for many years, but they could lose classified advertising to the Internet. Their businesses risk, in the words of a recent book on the subject, being "blown to bits."[8] Others may have margins so slender that a relatively small loss to the Internet could be disastrous. Many travel agents, for example, could easily be pushed out of business by the loss of a mere 3 to 5 percent of their market to the Web. In 1999, on-line penetration of the American travel market had already reached almost 2 percent.[9]

Many traditional consumer-goods companies have found it extraordinarily difficult to learn to compete with this disruptive technology. Few have been quite the high-profile disaster of the *Encyclopaedia Britannica*, but its fate is an awful warning of what could lie ahead. Founded in 1768 by three Scottish printers, it was bought in 1920 by Sears Roebuck and, after changing hands again, moved to Chicago. The company employed an aggressive direct-sales force which brilliantly persuaded middle-class families that an encyclopaedia held the key to their children's futures. By 1990, having flourished for more than two centuries, its market dominance seemed impregnable and its profitability assured.

Then came the CD-ROM. Microsoft's Encarta, an electronic encyclopedia launched in 1993, came free with many PCs. That made economic sense: it cost almost nothing to make an extra copy, compared with the $250 or so it cost to make a set of *Britannicas*. The encyclopaedia company fought back with its own CD-ROM. But first the company priced it high, to avoid undercutting its salesmen, then lowered the price and duly enraged them: how could selling a CD-ROM generate the $500-$600 of sales commission they expected? While company and distribution channel squabbled, sales of all hardbound encyclopaedias collapsed to a tenth of their 1990 level. The encyclopaedia was sold for a fraction of its book value and in October 1999 the new owners, taking a huge gamble, put the entire contents on the Internet, free.[10] Advertising, they hope, will recoup the cost – although, with a privately held company, they can veil the results.

The *Britannica* story sums up several of the difficulties that face "legacy" businesses. They are saddled with costly overheads: not just bricks and mortar, but sales forces, core competencies, and computer systems, that are designed to sell certain products in a certain way. Their on-line rivals start with nothing – a disadvantage in some respects, such as experience and branding, but a boon in other ways.

Even if established firms are willing to ditch some of these encumbrances, their profitability may depend crucially on the particular distribution channel they have always used. If a new, inexpensive distribution channel siphons sales from one over which they have greater control – as on-line stockbrokers have siphoned sales from conventional brokers – the older firms fear cannibalization (or rather, self-destruction). And if the new channel offers lower prices without necessarily generating lots more sales, it will destroy profits. Furthermore, established firms fear "channel conflict." Levi's, which experimented with selling jeans directly on-line, has run up against complaints from its other retail outlets.

Surmount these obstacles and another looms: the different culture, pay structure, and skills that seem to be needed to run an on-line operation well. Many companies try to buy in those skills or form alliances to acquire them. Wal-Mart, for instance, has teamed up with America Online and is setting up a separate Wal-Mart.com subsidiary which it plans to float off. Most of Europe's biggest Internet start-ups have not been free-standing, as in America, but spin-offs from established companies: firms such as Freeserve, an ISP create by Dixons, a British electrical-goods retailer, and subsequently sold to a subsidiary of France Telecom; or Egg, a bank set up by Britain's Prudential insurance company.

But the most successful transformations have been just that: examples of companies that have converted themselves largely into on-line businesses. One company to pull off this hugely risky step is Charles Schwab, America's biggest on-line retail stockbroker. Until 1996, it was mainly a discount broker, selling shares by telephone. Then it set up an on-line operation called eSchwab, which charged lower commissions. Confronted with the classic problems of cannibalization and channel conflict, it restructured itself into a single, predominantly on-line business.

Companies, though, have undergone wrenching transitions in their business model before. The task is tricky, but not impossible. IBM, after all, was built on the mainframe but survived the introduction of the PC to become largely a services organization. Once traditional retailers understand how to pull together their various distribution channels, they become formidable adversaries. While on-line retailers acquire a slew of new costs – better service, faster fulfilment, live telephone support for confused customers – traditional retailers are learning to get more mileage from their trusted brands and high-street shop fronts.

Thus Merrill Lynch has turned on eSchwab the mighty power of its brand, its research, and its analysts, together with the financial muscle to survive a commissions war. Barnes & Noble has found ways to use visits to its physical bookstores to recruit on-line customers. Department stores and discounters such as J. C. Penney, Nordstrom, and Sears have devised ingenious ways to use stores as distribution points, to accept returns and to offer advice and a friendly face. And catalogue retailers such as Land's End and L. L. Bean are building on their years of experience in fulfilling millions of small orders. Over time, on-line and off-line retailing will coalesce, creating new opportunities and services.

Consumers may ultimately feel more comfortable coming to e-commerce gradually, as an extension of the services their familiar retailers already provide, rather than branching out into the unknown and experimenting with a new supplier and a new technology at one and the same time. Traditional companies can move their loyal customers on-line for much less cost than a new on-line retailer must incur to buy shoppers' attention. Bricks may be an encumbrance, but many kinds of retailing will continue to need them.

Ordering and Distribution

Part of the potential of on-line retailing is that it offers a new way for customers to order and a new channel of distribution. Both cut costs. On-line ordering enables companies to shift work that they have done in the past to their customers, and learn more about them at the same time. On-line distribution enables companies to bypass traditional distribution networks and reach customers directly. But they must also surmount a challenge: the successful delivery of orders.

On-line ordering

One of the biggest cost savings that on-line commerce offers is in ordering. The customer fills in the form and the computer catches most mistakes. That saves large amounts of staff time. The savings will be greatest with the most complex purchases – taking out a loan, for example, or buying an airline ticket.

But customers are not so stupid that they fail to realize what is happening. They will expect something in return for doing the work. Companies can reward them in one of two ways. First, they can share some of the cost savings (although, astonishingly, some companies do the reverse and charge customers extra for ordering on-line). Second, they can offer customers a wider choice. Procter & Gamble has created a company called Reflect to develop on-line sales of beauty-care products. Its customers fill in details about their skin color, hair quality, and so on, and receive in exchange shampoos and cosmetics that have been made to their specifications. In the process, Reflect learns an enormous amount about its customers – and the customers receive a more distinctive product than they would find in a department store.

Reflect has even found a way to deal with the problem that plagues many on-line retailers: the abandoned shopping cart. Two times out of three, customers complete part of the ordering procedure, only to abandon it before the end.[11] Reflect can make up the lipstick the customer seemed to be about to order, and mail it free with an invitation to come back for more. No beautician, watching an uncertain buyer hovering near her department-store counter, could build on an uncompleted sale in quite that way.[12]

On-line payment

Ordering is simple, compared with paying. The currency of the Internet is the credit card, which accounts for about four out of five e-commerce transactions.[13] But many potential customers either dislike paying by credit card, or indeed do not have one. So security for payments is the single most essential requirement for electronic commerce to flourish.

When asked why they do not shop more on-line, customers often say that they are afraid of handing over their credit-card details. Airlines, for example, blame worries about credit-card fraud for their poor "look-to-book" ratio: of every hundred people who click on their Web sites, sometimes only one actually buys a ticket. Indeed, almost one-third of respondents to a survey by Forrester Research, an American consultancy, gave fear of credit-card fraud as their main reason for not buying tickets on-line.[14]

In fact, credit-card fraud is rife on the Internet. In the United States the main victims are retailers, which bear the risk when the cardholder is not present. A survey by Gartner, a consultancy, in July 2000 found that on-line retailers suffered twelve times as much fraud as off-line retailers. Fraud is especially common with purchases of products that can be downloaded, such as software and music, and that are easy to fence, such as consumer electronics. In addition, on-line credit-card transactions are more likely to be disputed than face-to-face ones. Even if a credit-card number is sent encrypted, making a transaction secure from the buyer's point of view, the seller cannot easily verify that the buyer really owns the card and authorizes the payment. If the buyer later queries the payment, and the seller has already despatched the goods, the seller loses twice over: the value of the goods and the value of the remittance to the credit-card company. Buyers can suffer too, of course. In Taiwan and El Salvador the cardholder is liable for the full amount of any fraudulent purchase.[15]

There are further disadvantages to payment by credit card – particularly if potential shoppers do not have them. In parts of Scandinavia and Germany, for instance, the convention is to pay on delivery for on-line orders. In some developing countries people dislike using credit cards because they leave an audit trail. And for small amounts – the purchase of a news clipping or a single song – a credit card is anyway an expensive and unwieldy option.

One answer to these obstacles may be smart cards. In September 1999 American Express launched its Blue Card, which carries a computer chip holding digital signatures unique to each holder. Consumers buy a reader that can be plugged into a PC, enter a PIN number, and use the card to make on-line purchases. Such cards are most likely to catch on if a single universal standard for reading them emerges.

Mobile-telephone companies hope to use their skills in billing for tiny amounts of talk time and messaging to bill customers for on-line purchases too. Sonera, a large Finnish telecommunications company, runs a pilot scheme in Stockholm to allow drivers to pay for car parking with their mobile telephones. Sonera has previously installed some fifty machines in Finland billed to mobile telephones, including car washes, juke boxes, golf-ball dispensers, and soft-drinks machines. But telephone companies may find it difficult to develop more general billing services to cover people who want to use their phones to order on-line via other ISPs, rather than buying directly from a selection of merchants offered by their telephone company.

That leaves electronic money. The initial products were flops. Users generally had to download special software, and the e-cash could be spent in only a few on-line stores. Some newer schemes enable people to earn points, like frequent-flier miles on airlines, by visiting sites (as with Beenz.com) or by buying (with a credit card) a gift certificate that friends can spend on-line (as with Flooz.com). Most have grown partly because they pay new customers $10 apiece. Some airlines themselves allow their frequent fliers to spend their accumulated miles on-line. That is huge potential spending power, given that there are perhaps three trillion unused miles, worth between $30 billion and $90 billion.[16]

As with so many areas of on-line retailing, these bright ideas will probably in time give way to schemes run by large companies that are already strong in an adjacent area. Thus First Data, a giant financial-processing firm, in alliance with Western Union, is building a scheme called MoneyZap that will allow customers to make on-line transactions directly from an ordinary bank account. Given that money is one of the easiest things to move on-line, it is a natural step for payment mechanisms to become seamlessly electronic too.

On-line distribution

When a product can be turned into digital bits and bytes, it can be delivered on-line. That has all sorts of advantages. First and most obviously, it cuts costs: no need for physical stores, for delivery trucks, for physical inventory. Second, it brings tax advantages. Levying taxes on digitally downloaded products will always be extremely difficult and require the (improbable) cooperation of customers. Both factors mean that the prices of products that can be delivered by digital download are likely to fall sharply compared with the prices of physical goods of all sorts. They mean, too, that new companies face lower entry costs than they would in the off-line world.

In addition, on-line distribution offers a way to reach new markets. It is an inexpensive way to distribute information globally. It offers an inexpensive way to test-market new products, such as software and games. And it presents a way for producers and customers to deal directly, cutting out any intermediary who was the distribution channel in the off-line world.

Here is where conventional businesses have most to fear. Whether record companies or stockbrokers, publishers or banks, their customer now has a less expensive and faster way to acquire their products. Airline tickets offer an example. For an airline to sell a ticket on-line through its Web site costs $1. Selling through a traditional travel agent pushes the cost up eight-fold.[17] A study by the OECD found that on-line distribution of products as varied as life insurance, airline tickets, and software costs between 50 percent and 99 percent less than using traditional distribution channels.[18]

So businesses are threatened if they made money by acting as toll gates, controlling a particular distribution route. Once that route is no longer needed, the tolls go too.

Luckily for slow-moving old-world businesses, on-line distribution is far from simple, so the tolls may survive a surprisingly long time in some industries. Writers might be able to bypass publishers and reach their audience directly – as Stephen King did when he put two instalments of his novel *The Plant* on-line in July 2000, offering to display the rest of the work if enough readers paid him directly $1 a chapter – but writers less famous will find that difficult. Even Mr King, whose previous on-line novella ran into technical problems with its encryption, eventually decided that his skills were more suited to dreaming up plots than devising on-line charging mechanisms. The balance of power between authors and publishers may shift, but publishers, like record companies and art galleries, will continue to perform for authors some functions that the artists would rather delegate. They will also, by acting as talent spotters and incubators, help to guide consumers through the otherwise bewildering superfluity of choice.

In other industries, intermediaries may also survive by redefining their role. Thus airlines, which have considerably more clout than a lone author, have begun to do something similar to Mr King: bypassing travel agents, and selling tickets directly to customers. The travel agents that survive will be the ones that add value by offering customers advice or other special services. In general, though, an intermediary makes money from the seller, not the buyer. Intermediaries are thus most likely to survive when they provide a service that sellers cannot efficiently undertake themselves.

Once consumers have more direct access to the product they want at a lower cost, the total market may start to expand. That has happened in stockbroking, the first consumer business to reach critical mass on-line. Here, the shift to the Internet has helped to create a whole new market of individual share traders. Improbably, perhaps, this shift occurred most dramatically in South Korea, where in the course of 1999 – 2000 share trading moved from a pursuit of the rich to a hobby for the masses. Stock-market turnover doubled in a year. In 2000, more than half of all trades were taking place on-line, and annualized turnover at one point reached 1,100 percent of market capitalization. Established brokers have not been squeezed out, but have undoubtedly been squeezed. Fighting to keep their business has forced them to drive down commissions to a minuscule 0.1 percent.[19]

Electronic distribution has drawbacks, though. Without good encryption, piracy is a constant threat; with it, distribution is more cumbersome. In addition, companies launching an on-line product may find it hard to pilot it in one region, as they would with a traditional launch. And reliability is a constant worry. Many new on-line launches collapse as the Web site is overwhelmed. When, in June 2000, Abbey National, a British bank, launched its on-line service Cahoot, the site collapsed within less than ninety minutes, swamped by the number of customers.[20] Eventually, reliability will improve. But until it does, many customers who have grown used to the reliability of the telephone and physical distribution channels will be deterred.

Distributing on-line orders

The trouble with placing an order on-line for anything that cannot be digitally delivered is the need to wait. And the need to be in when the delivery arrives. And the need to pay charges for postage, which may more than wipe out the savings that on-line purchasing promised.

. . . . .

When shoppers go to stores to buy, they do their own fulfilment: they select the goods from the shelves and they carry them home. When they buy on-line, the task needs to be done by somebody who is paid to undertake it. In this sense, the economics of order fulfilment are exactly the opposite of the economics of on-line ordering. In one, something the customer once did free has to be paid for; in the other, something once paid for is now done free by the customer.

The fulfilment problem overwhelmed many on-line retailers, new and old, at Christmas 1999. Toysrus.com was reduced to giving $100 vouchers to customers who did not receive their orders on time. Wal-Mart announced in the second week of December that it could no longer guarantee delivery of Web site orders by Christmas.

The experience taught many on-line companies that fulfilment matters to customers even more than price. But on-line retailers have tended to concentrate their energy on building and maintaining their Web sites, and to outsource fulfilment. Traditional retailers may also have problems with fulfilment: their delivery systems are designed to carry goods from warehouses to shops, not to people's homes. So order fulfilment is expensive, slow, and erratic for many companies. The survey by Shop.org and Boston Consulting Group found that, on average, fewer than nine out of ten on-line orders were fulfilled on time, and one in twenty was not fulfilled at all. One beneficiary has been UPS, which reckons to deliver about two-thirds of all goods ordered on-line, and has edged out Fedex, its main rival, which took too long to realize that document delivery, its main business, might be cannibalized by e-mail.[21]

Gradually, on-line retailers are creating their own distribution systems, either by developing central warehouses or by using bricks-and-mortar stores as distribution and pick-up points. Thus Amazon is building seven giant automated warehouses around the United States; Webvan, an on-line grocery store, has plans for twenty-one. But that is economic only for companies handling thousands of orders a day. Bain, a consultancy, calculates that it makes sense for a company to own its warehouse only when it handles at least fifteen thousand transactions a day. On its peak day, during Christmas 1999, Amazon handled sixty-one thousand. Few other on-line retailers are anywhere near that league.

Still, many small on-line companies try to compete on fulfilment speed. That means creating closer links between inventory, or stocks, and delivery. Pink Dot, a Californian store, aims to deliver groceries "in or around thirty minutes," using staff on bikes rather than a courier company. It hopes to defray the cost of local distribution centers by combining in them a physical store and a warehouse for on-line orders. It hopes that building stores will have an additional benefit: its brand will be visible in a way that a purely on-line retailer's brand is not.[22] Webvan is trying to guarantee delivery within a selected thirty-minute slot as a way of solving another problem: when people are out at work for long hours, who takes in the grocery delivery?

The question for e-tailers is whether the savings that arise from running a warehouse rather than a supermarket are enough to offset delivery costs. If not, are people willing to pay for the convenience of delivery? On-line shopping for physical goods is likely to grow more slowly than other kinds of on-line retailing. A price gap will open. A new divide between different kinds of retail commerce will thus emerge.

Informing and Marketing

· · · · ·

One of the frustrations experienced by Internet retailers is the enormous amount of window shopping. People look but do not buy. But while they look, they learn. And they use that knowledge to improve their bargaining power when making purchases. A classic example is the car market. In 1999, when fewer than 3 percent of car sales in the United States were made on-line, 40 percent of buyers used the Internet at some point, usually to compare prices.[23]

People spend a third of their time on-line looking for information. The Internet gives customers more information about products, and businesses more information about customers. It also creates a new opportunity for advertising and new ways of marketing. And it enables companies to watch more closely how their customers behave – if customers are prepared to accept the intrusion into their privacy.

. . . . .

Advertising

For many Web sites, on-line advertising is what is supposed to pay the bills. People like free content, and payment mechanisms are still crude – so young Internet companies (while spending lavishly in old media on advertising their on-line presence) plaster their sites with ads. Banners float across the top of screens; pop-up windows appear when you open up a Web page. In time, as more homes have broadband access to the Internet, advertisements may become more sophisticated – less like jazzy billboards and more like television commercials.

One potential merit, for the advertiser, is that the combination of the Internet's interactivity with its ability to gather data allows the customer to be targeted cheaply and effectively. Hunt with a search engine for "vacation," and a banner for a tour company may pop up. No need to show the ad to thousands of people who are not interested in going on vacation. No need to wonder how many people could have glanced at the ad: surveys measure visitors and the amount of time they spend at a particular Web site. No need, either, to ask whether, having glanced at the ad, people want to know more about the product. Advertisers can discover the "click-through" rate – the frequency with which people click on the banner to find out more (as fewer than 1 percent do).[24]

Targeting will grow more sophisticated. When people use a mobile telephone, it is possible to target advertising by location. Sunday, a mobile operator in Hong Kong, sends customers advertisements, either by voice or by text, matched not just to the customer's personal profile but to location, too.[25]

Some users may feel decidedly twitchy about handing over such data to advertisers. But people hand over lots of data knowingly to Internet sites, when they first register. Less knowingly, they deliver data through "cookies," small data files dumped by the site's server into the user's hard drive to track the user's on-line behavior. Companies use cookies to try to make sure that repeat visitors to a site see different advertisements. Increasingly, they also use them to build up a profile of the user, keeping track of the Web pages a user visits to infer his interests. Sometimes they use them in the same way as Amazon.com. It records the books a purchaser selects on each visit and then offers suggestions about other books that users with similar tastes have chosen.

All this data is potentially valuable: to the site, to advertisers, and to other companies that might want to buy it. In fact, tougher privacy laws (see Chapter 9) are restricting what companies can do with such information. And customers may grow irritable if, as Forrester Research predicts for 2004, marketing firms send an average of nine marketing e-mails a day to each household in the United States.[26]

However, advertisers have refined their view of on-line targeting as they have acquired more experience of it. It is, they frequently claim, more important on the Internet than with other media to know what users do than who they are. It is more effective to try to sell football boots to a consumer who is visiting a sports site, say, than to offer them to the same consumer on a visit to a banking site. That might sound disappointing for data-gatherers, but it also suggests that the segmentation of the Internet into innumerable sites could achieve a rough-and-ready targeting that is better than anything old-style market research could offer, even if it is a long way from the dreams of one-to-one advertising.[27]

For companies selling a product that is subject to advertising restrictions, such as medicines or cigarettes, the Internet is a potential boon. British American Tobacco (BAT) and R. J. Reynolds are both tentatively using the Internet to reach customers, although more as a way to exchange information with smokers than as a conventional marketing tool. The chairman of Brown and Williamson, an American division of BAT, uses the company's site for on-line chats; Philip Morris, another tobacco company, lists the ingredients of its various brands of cigarette; and R. J. Reynolds offers information on "smoking-awareness" programs.[28]

On-line advertising is a minute share of the total, and has been growing cheaper, thanks to the rush of new sites clamoring for ads. And new kinds of marketing have developed alongside them. For example, companies encourage consumers to do their proselytizing for them, using something called "viral marketing." To see it in action, note the inscription at the end of an e-mail from somebody who uses Microsoft's hotmail: each missive carries an ad, just as shoppers leaving Harrods carry the store's name on a swanky carrier bag.

. . . . .

Portals

Because the Internet is so vast and confusing, the greatest marketing power lies with a handful of frequently visited sites. One of the paradoxes of the Internet is that, with the consumer's choice unconstrained by government or channel capacity, the range of sites most people visit is even smaller than the number of television channels they watch. The top 1 percent of Internet sites capture more than half of the entire traffic on the Web, and the top 0.1 percent capture more than 30 percent.[29] So a company that wants to reach consumers needs a link with one of them.

These sites – AOL, Yahoo!, Microsoft sites, Lycos, Excite@Home – become portals, which attract huge numbers of users and then draw their attention to other, smaller sites. When a user clicks through and makes a purchase, the portal takes a cut. Click on, say, Alta Vista to buy a book from Amazon, and you will ensure that Alta Vista earns more from the purchase than the book's author. Such links are especially important for "pure-play" on-line retailers which, in 1999, drew 33 percent of their revenue from portal sites. By comparison, traditional bricks-and-clicks retailers, with their well established brands, drew only 12 percent of revenue from portal sites. A little known company seems to need the clout of a portal at a well known site to bring in custom.[30]

This arrangement has encouraged all sorts of sites to position themselves as portals. Despite the Internet's capacity for destroying some kinds of middleman, a whole new kind of intermediary has thus been created. The main service that many of these new intermediaries provide is (in theory) trust, an essential commodity in a world in which buyer and seller never meet face to face. So companies that already have strong brands and lots of information about their customers – such as banks – hope to use those advantages to steer people to other sites, and to make money from their role as trusted guide.

Customers as clubs

Take the combination of a brand and trust, and companies have the basis for trying to make their customers feel part of an exclusive club.

The aim is for a company to avoid becoming a small-margin peddler of commodity goods and services by using the information it has on its customers to sell them other products that they might not at first associate with that brand. Take the airlines' frequent-flier clubs. Their members are united, not by a common interest, but by the fact that they have jobs that require them to spend many hours in the seats of Continental or Air France. But the airlines strive to raise their margins by selling these regular passengers everything from luxurious holidays to bottles of wine.

New communications make it easier and far less expensive to build and manage such clubs. They can be global: physical and national barriers are less relevant. It becomes cheaper to keep a database and to run an after-sales service: indeed, many "clubs" will grow out of the trend for manufacturers to slither into service provision as they find new ways to keep track of the physical products they have sold, sending existing customers news of upgrades and new products.

It also becomes easier and cheaper to interact with customers, allowing the club members to share their thoughts on the service they have bought. An important effect of Internet-related technologies is to shift the emphasis in companies away from recruiting new customers and toward building deeper relations with existing ones. It is less expensive to sell the same product to somebody who has already bought it once than to build new markets. Many companies will find that their customer "club" is their most valuable source of product testing, of recommendations that bring in new customers, and of market research.

Personal pricing

For shoppers, one of the empowering effects of the Internet is to provide information on prices. Trailing around the shops or telephoning through the Yellow Pages is one way to compare prices; less tedious by far is to click on a site that offers "shopping bots" – intelligent software that scans the catalogues of on-line merchants. For retailers, though, a more radical effect of the Internet could be that it will enable them to discriminate more carefully in the prices they charge different customers.

In theory, price-comparison sites enable shoppers to compare prices, products, and availability. Customers can download a piece of software that automatically tells them the best available price for the product they want. About to buy a CD from CDNow? A box pops up and shows where the same disk is available for less. Such sites will increasingly be available on Internet-enabled mobile telephones, so that customers can use them while they shop rather than before they leave the house.

In fact, many of these services include only sites that have signed up to allow their information to be displayed. Many, indeed, earn a commission when a buyer clicks on an item that its bot has tracked down.

Some sites also compare products. They may carry reviews by journalists, or individuals may swap their own experience. In time, this combination of an ability to search and compare with a bulletin board for communal customer experience, could be one of the most attractive and powerful tools of on-line shopping. For the moment, it offers a lesson on intermediaries that applies as much in the on-line world as the off-line: the advice they give the consumer depends heavily on who is paying the bill.

While the new availability of information on pricing will influence many on-line retailers, it is clearly not the only force at work. Indeed, one of the paradoxes of the Internet is that prices vary widely – sometimes more widely than in the high street or shopping mall. One study found that the prices for identical books and CDs at different Internet retailers varied by as much as 50 percent, with the differences averaging 33 percent for books and 25 percent for CDs.[31] Another study found that airline tickets from on-line travel agents varied by an average of 28 percent.

The reason for such price dispersal seems to be that consumers are keener to buy from a trusted brand than to save money by paying the lowest possible price on-line. Yet another study found that retailers with well known brands, such as Amazon.com, could charge a premium of 7 percent to 12 percent over more obscure retailers.[32] On the high street, the buyer can check whether a new store looks sleazy or the shopkeeper seems honest. On-line, brand and trust seem to matter more.

Taking advantage of such attitudes, some retailers will see a way to fight back against the downward pressure on prices. Already, Internet retailers change their prices more frequently, and by smaller amounts, than off-line retailers do.[33] They may well use the flexibility of the Internet for greater price discrimination. Customers will be offered a wider choice of services. Express? Super-express? Ordinary? Repeat customers can be spotted easily and rewarded; customers in a hurry can be charged a little more than those who are clearly bargain-hunting. So customers who might willingly pay a bit more will do so, while unprofitable customers can be identified and charged a higher price. Consumers, used to paying different prices for some products (like cars) but not for others, may feel uneasy about this change. But, like cunning shopkeepers in an Arabian bazaar, on-line retailers will try to spot each individual's pricing point, and sell at just that mark.[34]

A Marketplace

. . . . .

As well as being a distribution channel and a source of information, the Internet is a marketplace in its own right, offering innovative ways to set prices. It is the perfect base for an auction. Not only does it allow many buyers and sellers to come together at the same time, giving the depth and breadth that a good market requires, but its software also allows an auction to be structured in all manner of fancy ways, while its low overheads hold down commission charges. Its infinite shelf space enables many niche retailers to flourish. In addition, a more individualized relationship between buyer and seller is possible on-line, so that people can, in effect, design exactly the product they want.

Auctions and aggregation

In a sense, the Internet permits a return to the world of two centuries ago, when most trading took place in markets or bazaars and prices moved constantly by small amounts in response to supply and demand. So far, eBay is the largest site at which consumers can sell things to one another. Mostly, they seem to sell Beanie babies and the like.

Priceline offers a different kind of auction – the Dutch sort, where bidders say what they are willing to pay (in this case, for airline flights) and wait to see whether anyone will supply them at such prices. However, because the company refuses to publish prices, it lacks the transparency of a good marketplace.[35]

A third model is that of NexTag, which allows buyers not just to submit bids but also (unlike Priceline) to come back and change them. Sellers can accept or reject them, or counterbid. The result is more like a bazaar – or indeed, a stock exchange.

A number of companies have created sites that gather consumers together and marshal their buying power to drive down prices, with mixed results. At the end of 2000. Letsbuyit, a Swedish company, which claimed to be the first pan-European site of its kind was one of many fashionable business-to-consumer companies in financial trouble.[36]

Of all the markets that consumers use on-line, none is more sophisticated than those that deal with money. Several companies offer direct-access trading platforms for investors buying and selling equities. They enable private investors to bypass a broker and send their orders to whichever marketplace they choose for execution.[37] Such markets may include electronic communications networks (ECNs): automated exchanges, privately owned, that can automatically match buyers and sellers without the need for human intermediaries (although for the moment, they link only stockbrokers, not their clients). By the beginning of 2000 these markets traded the equivalent of 12 percent of Nasdaq's volume. Financial regulators view them with suspicion, however: despite their low trading costs and high efficiency, they lack the liquidity of a conventional stock exchange. Eventually, though, individuals may be able to trade automatically not only stocks but foreign currencies and other financial instruments, on markets that would be more volatile than those that specialists alone can use, but cheaper to deal on.

The Internet's many new intermediaries, with their low overheads and broad reach, do not need to store stock. They merely bring together buyers and sellers. They benefit enormously from economies of scale: the more buyers come, the more sellers will be attracted and vice versa. They will never be used for all consumer purchases, or even for most – but for retail finance, for expensive items such as white goods and cars, and for services such as airline and theater tickets that have a limited shelf life, they may become one of the most popular ways to buy.

Niches

More valuable to shoppers than even its scope for cutting prices may be the Internet's ability to enhance choice. The unlimited shelf space of the Internet makes it the ideal vehicle for niche companies selling obscure products. Adam Smith's observation of two centuries ago that the degree of specialization is related to market size is daily demonstrated on-line. With a valuable product that is hard to find in a local high street, but that can be shipped around the world, a tiny trader can reach a global market.

Take three British examples. James Barber Tobacconists, a 133-year-old Yorkshire family business that sells collectors' smoking pipes, sells in seventy countries and finds that 60 percent of its sales are electronic. Its Web site, built on a whim by a local policeman, generates about thirty order inquiries a day. The company has overtaken Harrods, Britain's premier department store, as the largest seller of Dunhill's exclusive pipes. Meanwhile David Marshall, in rural Cambridgeshire, makes instruments such as bagpipes and crumhorns. His Web site, built by a friend, generates commissions with an average value of almost $1,000 from countries as far apart as Sweden, Australia, Italy, and the United States. A friend put up a Web site as a surprise for Tony Moss, a retired college lecturer who runs Lindisfarne Sundials of Northumberland. Mr Moss, who had previously tried in vain to drum up foreign business – by, for example, buying a center-spread in a leading American horticultural magazine – promptly found himself "inundated" with electronic inquiries and turning away surplus commissions.[38]

Similar tiny businesses are starting to flourish in many countries by using the Internet to reach niche markets. Their success will depend partly on how effectively search engines can help customers to locate exactly what they need. They must also deal with some of the problems that face more conventional on-line retailers: shipping charges, customs duties, payment, and fraud. But customers for these niche products will be more interested in finding exactly what they want than in using the Internet as a way to buy cheaply.

. . . . .

Mass customization

Among the many ways the Internet empowers consumers, one is to enable them to choose more precisely the product they want. As with Procter & Gamble's Reflect, shoppers can assemble a product tailor-made to their tastes.

Mattel, an American toy maker, allows American customers to design their perfect Barbie doll, with hair, eyes, personality, and name all selected from innumerable permutations. Orders travel electronically to Mattel's production line in China, to be assembled and shipped back to the United States. DBS Oegland, a Norwegian maker of bicycles, invites visitors to design their own version of the Intruder, the company's top-of-the-range model. Concentrating on variations on a single, upmarket product seems to work well: individuals are willing to pay a bit extra for a product made to their own specifications.[39]

But companies offering what Joseph Pine, who wrote a book on the subject, calls "mass customization,"[40] quickly learn that customers need sites that steer them simply and cleanly through the choices, rather than overwhelming them with variety. Companies and customers jointly need to learn to work together in new ways. And companies, as the next chapter explains, need to put the relationship with the customer ahead of everything else.

A Global Market

. . . . .

Electronic commerce, at the start of the new century, is still mainly an American business. One of the main opportunities, though, is to create a global market. Already, electronic commerce outside the United States is more export-oriented than sales by American firms. In the United States, export sales account for only about 10 percent of e-commerce revenues. In Asia/Pacific, it averages 38 percent; in Latin America, 79 percent; in Canada, 83 percent.[41] One of the greatest opportunities that electronic commerce creates is to build new export markets, especially for developing countries.

American dominance

For the moment, though, America's dominance of e-commerce is indisputable, and growing. Measuring electronic commerce is difficult, but one of the best ways to do so is by looking for secure servers – software for encrypted transmission, which creates a secure end-to-end link across which credit-card details and the like can be sent to sites designated by the initials "htts" instead of the usual "http." Of the sixty-seven thousand secure servers in the rich countries of the OECD in March 2000, 70 percent were in the United States. And America was pulling away from the pack. In the ten months to April 2000 the United States added ten times as many secure servers per head as the whole of the rest of the OECD put together.[42] [Fig 5-2]

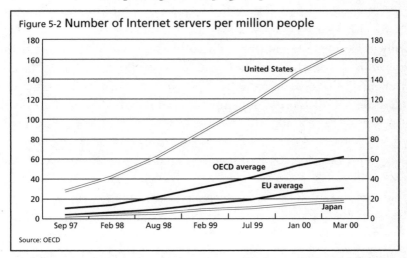

Figure 5-2 **Number of Internet servers per million people**

Source: OECD

Not surprisingly, American on-line retailers have made hefty inroads into markets in sleepier parts of the world. In Western Europe, they grabbed 20 percent of the market by the dawn of 2000; in Asia, 14 percent.[43] Companies such as Dell Computer, Amazon, and eBay are all expanding in Europe and Asia, trying to stake out on-line territory before local competitors wake up to the opportunities.

. . . . .

Dealing with diversity

In the past, American companies have unified commerce by investing in many different countries. Now, businesses such as Dell and Amazon are helping to break down insular European attitudes to e-commerce. Yet the sheer diversity of Europe and Asia – cultural, linguistic, socio-economic, regulatory – will inhibit cross-border growth. To reach 70 percent of Europeans requires content in five languages (although some are prepared, for the moment, to rub along in English).

Other differences are more surprising. The use of credit cards is much lower in Europe. In the Netherlands and Belgium, for instance, roughly half of on-line payments are made with something other than a credit card. That requires a greater mix of off-line services (for pick-up and distribution) with on-line than has so far been common in America. In addition, the quality of local delivery services varies widely. One study in Germany found that 80 percent of goods ordered on-line and sent from abroad arrived with "significant" additional charges.[44] More fundamentally, in many countries people are simply not used to the whole idea of "distance" shopping, so natural in the United States with its Sears Roebuck tradition. In most large European countries, fewer than 20 percent of people have bought books or CDs by mail order.[45] Some countries, such as Portugal, have hardly any experience of mail order at all.

Such barriers will take time to break down. Others, such as the logistics of shipping goods across borders, will hamper cross-border retail commerce for many years. Some products will inevitably be more readily traded across borders than others. Thus the market for fresh food will always be mainly local, even if the market for CDs is global. Even when companies sell on-line, some will want to sell only to a local market – to avoid price comparison, regulatory difficulties, or the sheer nuisance of cross-border trade, or to cater for local tastes. They will find it increasingly easy to steer unwanted shoppers to their own local on-line supplier.

Already, companies are learning to tailor their Web sites to different markets. QXL.com, Europe's answer to eBay, has a string of on-line auction sites across Europe, specifically designed to cope with different languages and currencies. Or take ConSors, one of Europe's largest Internet brokers, and the subsidiary of a German company, Schmidt

Bank. To move into France, the broker set up an office in Paris with an all-French staff and offered support services entirely in French. The company's strategy is to create a local product for each country. As a result, the Internet becomes a way to spread global business models and products, designed for local demand.

So the Internet will offer new ways to meet local demand. But it will also encourage cross-border trade and investment. Europe will be the main testing ground.

Many companies see the Internet as a relatively inexpensive way to move into foreign markets, reaching the most adventurous and best-educated consumers. So America's Citibank provides retail banking in a number of European countries but has used the Internet to widen its reach, creating what it calls a "branch in the sky." Deutsche Bank plans an Internet-only bank in India. In Europe, cross-border banking is helped by the fact that a financial-services organization with a license to operate in one EU country can offer services in another.[46]

In addition, the coming of the euro, a common currency for eleven countries – including France and Germany, two of the largest – makes it easier for people to compare prices across the continent. It will thus become harder to sell products for widely varying amounts in different countries. Identical cars have long been more expensive in Britain than in neighboring European countries, for example. Several Web sites quickly sprang up to help British customers buy cars in Belgium or the Netherlands. Such cross-border "grey-market" trade is likely to surge through other goods with big price differences, such as perfumes and pharmaceuticals.

In Europe, the Internet, together with the single-standard GSM mobile telephone, may give the continent a degree of unity it has never previously enjoyed. By the end of the decade, one of the main consequences of the Internet may be to bind together the disparate nations of Europe in commerce and communication, thus turning the ramshackle European Union into something closer to a real common market.

For consumers everywhere, electronic commerce will eventually bring empowerment: to search, to bargain, to specify; to have what they want, when and where they want it. That will be a delight for American consumers, of course. But for consumers in many other parts

. . . . .

of the world, where choice is narrower and markets more rigged in favor of producers, it will be a revolution.

Corporate Commerce and Company Structure

"Most successful Internet companies are really in retailing or computers or finance or media or some other service or manufacturing business. The Internet is just the tool they use to do business incredibly efficiently and quickly. It flows through the whole operation just like electricity. A retailer doesn't say to customers, 'come to my store because I am wired up to electricity or because I keep the air conditioning on.'"[1]

Speaking at the World Economic Forum in Davos in January 2000, Michael Dell, founder of Dell Computer, summed up one of the key points about the effect of new communications on companies. They are not, of course, confined to dotcoms. Their most important impact – indeed, arguably the most important impact of the Internet in the first decade of this century – will be on existing companies, revolutionizing the ways they do business. They will alter the ways companies interact with one another and with their staff, encouraging them to look outward rather than inward. They will affect the way businesses reach their customers and process transactions. They will influence inventory control and, indeed, the whole way companies manage their supply chains. So profound are the changes under way that they promise nothing less than a transformation of the nature of the company itself.

Companies are the first to be affected partly because they are the best prepared for change. Moreover, because they make greater use of communications, they are likely to value their connections more and be ready to pay more for them. Because businesses tend to be clustered in districts, their connections can be upgraded relatively inexpensively. Not surprisingly, business districts have been the first recipients of high-quality fiber-optic networks that make possible fast and easy Internet connections.

In addition, the changes in communication chime with other trends in the corporate world and reinforce them. One such trend is globalization, whereby world trade and investment comprise a rising share of total economic activity. Another is the booming market for ideas and knowledge: increasingly, creativity and the ability to process information are key competitive tools. Businesses must also respond to the incessant pressure to control costs. With the return to an era of low inflation, reducing costs has become an essential survival tactic for many companies. Related to this is the impetus to cut layers – both internal (middle managers, for instance) and external (intermediaries). Flexible employment patterns, designed to acquire the maximum expertise at minimum cost, create different relations between companies and their workers and require flexible communications.

Adept use of communications offers a chance to create new business models and find entirely new ways of doing things. Unlike some of the large investments that companies have previously made in overhauling their computer networks, this one is different. It does not merely streamline what happens within a company. It changes the way companies look outwards. Openness becomes a corporate strategy as companies allow customers and suppliers unprecedented access to their databases, staff, and inner workings.[2] New opportunities for building alliances and relationships demand greater trust in corporate life. Learning to manage these new relationships in novel ways will clearly be an important – perhaps the most important – competitive advantage for businesses in the new century.

The Pace of Change

It is a truism to say that the world in which companies operate is changing faster than ever before. The Internet enables that change, sometimes making possible innovations that were long desirable but not previously feasible. It spurs innovation: ideas and knowledge flow more freely, and because every company can see that conventional businesses will alter, the incentive to apply knowledge is greater. It spurs competition: by lowering some barriers to entry and by extending the size of markets that were once protected by technology or geography, it allows new e-businesses to spring up and rival established firms. And those companies that have already moved their core activities on-line have an enormous incentive to push their suppliers and customers to become part of their network, creating a vast ripple effect.

Thanks, perhaps, to all the furore of the dotcom boom, companies in most of the rich world have begun to grasp the Internet's importance. Most senior managers now believe that it will eventually affect their business. Back-of-an-envelope estimates of the potential cost savings, such as those by Goldman Sachs, are a further encouragement. [Fig 6-1] Not surprisingly, the rate at which companies exploit the

---

Figure 6-1 **Potential business-to-business savings**

Industry	Potential cost savings, %	Industry	Potential cost savings, %
Electronic components	29-39	Chemicals	10
Machining (metals)	22	Maintenance, repair and overhaul	10
Forest products	15-25	Communications/bandwidth	5-15
Freight transport	15-20	Oil and gas	5-15
Life science	12-19	Paper	10
Computing	11-20	Health care	5
Media and advertising	10-15	Food ingredients	3-5
Aerospace machining	11	Coal	2
Steel	11		

Source: Goldman Sachs

Internet has been rising rapidly. Surveys in Denmark and Finland, two countries at the sharp end of Internet adoption, found that the proportion of firms with more than twenty employees using the Internet to place orders rose from about 15 percent in 1997 to more than half in 1999, while the proportion receiving orders on-line shot from 7 percent in 1997 to about 40 percent two years later.[3]

In the United States, the Internet was at first used mainly as a marketing channel – a reflection of the hype attached to business-to-consumer applications in the second half of the 1990s. In 2000, though, many companies began to apply it to managing procurement and the supply chain, realizing that it was a far less expensive tool than earlier, proprietary electronic-data networks. This shift is extremely important. As long as the Internet is used only as a distribution channel, to sell to retail customers, it can to some extent be seen as a rival to the telephone, the mail, or a store. Once it becomes integrated into internal operations and the supply chain, companies begin to see the need to make larger changes in business structure. They use it for procurement, and they also tend to develop new links with one another. [Fig 6-2]

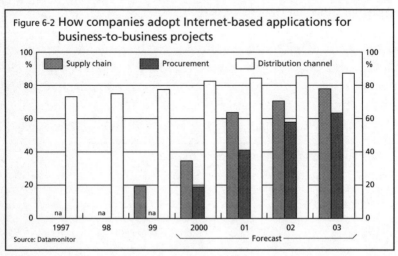

Figure 6-2 How companies adopt Internet-based applications for business-to-business projects

Supply chain   Procurement   Distribution channel

Source: Datamonitor

In time, the Internet has the potential to transform business processes as fundamentally as electricity did, perhaps more so. But precisely because that transformation is so extensive, it will take time. Not until Henry Ford spotted the carcasses revolving in Chicago slaughterhouses, and made the imaginative leap from fixed to moving production lines, did the full benefits of electricity feed through to mass production, and thus into raising living standards across the world. The equivalent change in this century may take just as long, because it requires more than a mere rearrangement of machinery. It calls for new ways to structure a business and to work with other companies. It needs different business processes and workers with different ways of thinking. Moreover, these innovations will differ from company to company, with the result that firms may not be able to learn from one another's experiences, as they could with the introduction of electricity or even the telephone. So, even though companies will be the Internet's pioneers, their gains may take at least two decades to work through.

Information and Knowledge

. . . . .

Before most corporate managers knew what the Internet was, they had begun to see that handling knowledge and information would be one of the keys to corporate success. The reason is simple. The cycle of innovation is faster and successful innovation is a bigger driver of corporate growth. The arrival of the Internet has given companies both a new tool for managing knowledge – enabling it to be codified, searched, and shared – and a new incentive to do so. One result has been a host of experiments aimed at putting internal information on-line, from routine document-processing to designs and technical drawings. Another has been an attempt to find new uses for corporate knowledge, by trading it or by using it collaboratively across the supply chain. Both are particularly important in increasing the productivity of services – the services companies use in-house and those they sell to other customers – because so many services involve the processing of information or the diffusion of knowledge.

Sharing and processing routine information

Many of the early efficiency gains companies have made arise from using the Internet as a way to handle routine transactions. Simple but ubiquitous tasks such as order-processing and after-sales service can be automated relatively cheaply.

As companies sell increasingly sophisticated products, the costs of these simple tasks tends to rise; in some cases they may account for 10 percent of operating costs.[4] Thus shifting order-processing on-line cuts costs. But it also increases accuracy. It is harder to put wrong information into an automated system than to give it to a human being. Cisco Systems replaced ordering by telephone, fax, and even e-mail with an automated procedure that checked an order's internal consistency and matched order, receipt, and invoice. The result was to cut the number of orders that had to be reworked from around 25 percent to 2 percent.[5]

Similar savings arise from moving purchasing (of stationery or other office supplies, for example) on-line because it becomes possible to apply systems to relatively small orders. It is also easier to make sure that everybody complies with company purchasing policy. This applies not just to orders of office supplies. Bell South, a big American telephone company, cut the approval time for expense reports from three weeks to two days.[6] And when GloMediX, a European company that specializes in electronic ordering of hospital supplies, looked at how much hospitals spent on items such as syringes and drugs, it calculated that it could cut the cost of a typical order from 145 euros ($130) to a mere five euros.[7]

In many firms, customer-service departments gobble up at least a tenth of operating costs. The Internet offers an alternative. A common approach is to allow customers access to information that they would previously have telephoned to ask for. Thus customers can see their bank statement on-line, interrogate a database (to check a flight time, for instance), or track the progress of an order through the system. Both Federal Express and UPS enable customers to check exactly where their package is, day or night.

Cisco Systems offers a classic example of the substitution of on-line support for expensive customer-service departments. Faced with the need to provide customers with continuous and expensive technical back-up, and with a daily flood of inquiries, Cisco decided to move as much of the task as possible on to a Web site. Not only did the site become a place where customers shared their experiences with one another and with Cisco, but the company also estimates that it eliminated about 250,000 telephone calls and saved the equivalent of 17 percent of operating costs.[8] Switching to on-line customer support can cause fury if badly handled, however. Plenty of surveys find that companies respond more slowly to e-mailed inquiries than to telephone calls – if at all.

A parallel opportunity exists for dealing with inquiries from staff about pay slips, pensions, conditions of employment, and so on, many of which are handled by human-resources departments. Again, companies are starting to make such information more accessible to staff on-line. Oracle, for example, allows employees to update their own pension arrangements.

Companies often spend large sums updating and distributing paper reference books, such as internal telephone directories and staff manuals. All of these can now be stored and altered on-line. When Renault, a French car maker, bought into Nissan, a struggling Japanese car company, it adopted English as the common language, then wrote the company dictionary of English on a spreadsheet and posted it on-line so that it could be swiftly searched or updated.[9] Meanwhile, repair people working for Centrica, a British company that specializes in servicing a vast array of domestic appliances, use a mobile link to the Internet to consult the pile of manuals that they would otherwise have to carry in their vans.

Increasingly, companies save money by automating entire processes, rather than parts of them. When customers send a stock-buying order to Charles Schwab Europe, a leading on-line brokerage, the order is routed to a system that automatically locates the best price, completes the deal, and sends notification back to the user on-line. The whole process, which once took a couple of minutes, now takes ten seconds. No paper-shuffling, no human intervention – apart from the customer, who drives the whole process with a click of a mouse.[10]

Managing projects

For managers faced with running a large and complicated project, the Internet's ability to transmit and search for information is a godsend. It enormously reduces paperwork, limits the scope for error, and enables companies with many different skills to work together seamlessly.

A building project is a prime instance of the need for collaboration – not just within a company, but externally, with dozens of businesses (architects, engineers, material suppliers) over periods of months or years. Without the Internet, each project entails thousands of transactions, all recorded on paper. A typical $100 million building project generates 150,000 separate documents: technical drawings, legal contracts, purchase orders, requests for information, and schedules. Project managers build warehouses just to store them. Federal Express reputedly earned revenues of $500 million in 1999 solely from shipping blueprints across America.

To cut these costs, some companies now create a Web site to which everyone involved in a project, from the architect to the carpenters, has access in order to check blueprints and orders, change specifications, and agree delivery dates. Moreover, everything from due dates to material specifications is permanently recorded, creating an audit trail that is a particular advantage in a notoriously litigious industry. Swinerton & Walberg Builders, a large American contractor, says that using such a Web site has reduced by two-thirds the time needed to deal with requests for information.[11]

A similar process enables Davis & Co, a London law firm, to coordinate teams of lawyers working on due diligence in large mergers or takeovers. A global takeover readily generates up to thirty thousand pieces of paper and involves many different disciplines, from accountants to auditors. The company has created a secure Web site on which both clients and specialists can monitor progress or hold an impromptu discussion of a document using an on-line "whiteboard" to mark up amendments. One such project coordinated fifty lawyers, fifty accountants, and fifty due-diligence specialists working in twelve cities across nine countries, from Australia to Kazakhstan.[12]

Knowledge and innovation

The Internet also transforms the way in which companies can share knowledge and ideas, either internally or in collaboration with partners. Indeed, collaborative working was the Internet's original function back in the days of ARPANET, when it enabled defense and communications boffins in different universities and government departments to work together on the same project.[13]

Sharing information and knowledge becomes easier at every level. Many companies now have a database of knowledge similar to the one operated by Accenture (as Andersen Consulting now calls itself). The Knowledge Exchange, a vast on-line compendium of the firm's accumulated wisdom, is available to Accenture people anywhere in the world and at any time. Other companies use the Internet to enable employees to scour the organization for advice or best practice.

For example, BP, a British oil company, has developed an intranet called Virtual Team Network that links personnel on oil-drilling rigs around the world. If a drill bit develops a fault, a rig worker can log on to the network and put out a worldwide request for help to mend it. Xerox, an office-machinery giant, has built Eureka, an intranet linked to a database that enables its twenty-three thousand service staff to share tips on repairing the company's copiers. (The reward for a good tip is not cash but the admiration and gratitude of fellow workers.) And Buckman Laboratories, an American specialty chemicals manufacturer, has an intranet called K'Netix to allow its sales people to tell the company's research and development staff about demand from customers for new products. The idea is to build on the fashionable Internet concept of "communities of interest." A product manager might even be able to start work on developing a suggestion straight away, without waiting for the word from head office.[14]

As such examples show, the Internet makes it easier to handle information flexibly in a company. It also directly encourages innovation, which feeds on shared knowledge, in a number of ways. For one thing, information and communications technologies (ICT) themselves have the highest rate of innovation of any industry, when measured by patent applications.[15] That swiftly spreads into innovation in other industries, especially services, which are increasingly built around software, computers, communications networks, and data banks. All told, about a third of the research and development carried out in the services sector is related to information technology.[16]

Another effect is to make it easier for companies to monitor what their rivals are doing and to watch other companies around the world and in different industries. This monitoring has become an important part of the process of innovation. New ideas applied by one industry in one market, such as mass customization, are quickly taken up and applied by others in other sectors.

In addition, it is easier for companies to "buy in" innovation, rather than "making" it all in-house. Sometimes, this involves strategic alliances between companies in different fields; sometimes, the purchase of business services that incorporate new ideas. In a study of the links between innovation and economic growth, the OECD found that the demand for business services was growing rapidly and that services such as consultancy, training, research and development, and computing all played an important part in diffusing new ideas.[17] These service companies not only help other companies to understand how to make the best use of new communications; they are often big users of information technology themselves.

Finally, by allowing many different specialists to work simultaneously on a new project, the Internet has encouraged research and development to move from corporate laboratories to business units. In some cases, these units have been separate young businesses in which larger firms hold stakes (companies such as Intel, Cisco, and Microsoft routinely act as venture capitalists to small start-up firms). In others, a former R&D laboratory becomes a key part of a company's business strategy. Boeing's Phantom works, once the R&D heart of McDonnell Douglas, is now a "tool for sucking knowledge out of different parts of the huge company and squirting it into a business plan."[18]

A new tool for talking

Transferring and sharing information in all of these ways is becoming easier, thanks to an important technical advance: the development of a new language to describe Web pages, called extensible mark-up language, or XML. This may turn out to be the key breakthrough in enabling companies to use the Internet for a vast range of tasks.

Big companies have long had proprietary electronic networks. One, electronic data interchange (EDI), links companies to their suppliers. But these networks are expensive to install and cumbersome to use. On the other hand, the Internet also has weaknesses: if one company sends an invoice electronically to another, the receiving company's system will not instantly recognize what it is. A human being will have to type the information in again, or change the format.

XML is a way of describing the content of a Web page in terms of the type of data it contains, rather than the way it looks. So it allows another company's system to understand that the material it has been sent should not just be set in capital letters, or small type, but is an invoice or a purchase order. The result will be to allow information to pass from one company's system to another's without the need for human intervention.

However, companies still need to agree on rules or standards to ensure that invoices or purchasing orders use common terminologies. A fishing fleet sending data to fishmongers, for example, would need first to agree on whether the measurements it used for particular kinds of fish were to be stored by weight or by size. Such standards, crucial for commercial interaction, will emerge in the next few years. Once they do, XML will become the core of electronic commerce.

Buying and selling

· · · · ·

Innovation these days rarely means just building or buying a better gadget. It means transforming business processes. For most companies, the road-to-Damascus moment comes when they realize that the Internet offers a way to reorient the entire business so that customers at one end of the chain and suppliers at the other are linked seamlessly together. At that point, the company often requires a wrenching reorientation. No longer is it designed to keep the customer out of the machinery of the business; instead, the customer sees the internal workings with greater clarity than ever before. No longer does supply push production through the firm; instead, demand from customers pulls it. Once that occurs, companies often think more carefully about what they actually do, and why (or whether) it is necessary for them to stand between customers and suppliers in this way.

Purchasing

Many companies are finding that buying on-line is one of the surest ways to make savings. Purchasing falls into two rough-and-ready categories: direct materials that go into end-products (such as parts or chemical feedstocks), and indirect materials, which may be anything from carpets to lubricants to hotel accommodation for traveling staff. The arrival of the Internet changes both kinds of buying, but many of the easiest savings are in the second category.

Often, the first kind of buying has been on-line for years, although the systems have been proprietary, inflexible, and expensive. Industries such as car making and food manufacture – in which suppliers frequently replenish their inventory – long used EDI to keep suppliers in touch with demand. Most EDI networks were imposed on suppliers by the buying company – a supermarket, say – and required a dedicated electronic connection, a hugely expensive investment for a small firm. Because they were based on a proprietary technology, they also locked companies and suppliers together. Installing the same facility on the Internet, or building a corporate extranet (a piece of the Internet walled off for designated users only), is not only less expensive,[19] it is also

. . . . .

based on open standards so that corporate purchasers may enjoy a more flexible relationship with suppliers.

The second kind of buying has usually floated free. Buying toner for the photocopier can sometimes be done on the say-so of the local office manager, and sometimes needs a sign-off from the buying department. One way, it tends to be extravagant and hard to track; the other, expensive to manage and infuriatingly slow. When the local manager places the order, a big company may use hundreds of suppliers; "rogue" or unauthorized purchases proliferate; and company purchasing policy becomes impossible to enforce. When the buying department does the ordering, the resulting bureaucracy may cost more to run than the value of the purchases themselves.

One of the fastest ways for companies to save money is therefore to use the Internet to try to bring indirect purchases under better control. That involves negotiating centrally with suppliers, drawing up a single catalogue, and insisting that staff either buy from that or explain why they want something different. Some of the biggest savings often come from cutting out those "rogue" purchases. In addition, the result is often to bring down the number of suppliers. When British Airways moves procurement on-line, for example, as it was doing in the course of 2000, it expects its suppliers to dwindle from fourteen thousand to about two thousand.[20]

One area in which this kind of operation has been especially successful is in the purchasing of office supplies. Big companies need stationery and furniture, but often these purchases have been scattered among many departments, preventing the firm from achieving the economies of bulk buying and also making it harder to impose a single corporate policy (on letterheads or business cards, for instance). By consolidating these purchases, companies can make savings.

Some groups of companies are beginning to pool their indirect purchases. Twelve large American corporations from an assortment of non-competing industries – including Bethlehem Steel, Kellogg, and Prudential Insurance – have pooled their buying power to create a single purchasing consortium for requirements such as energy, advertising, and marketing. The venture, managed by a subsidiary of Electronic Data Systems called CoNext, aims to aggregate demands and then invite suppliers to bid for contracts.[21]

Another approach is to hand the whole task over to a specialist. One example: an American firm called ImageX allows customers to set up on-line templates for business cards, stationery, and brochures. Employees can customize the templates and order the goods through a password-protected Web site. The work is then subcontracted to a network of just over a hundred printers. ImageX sends a digital file straight into the printer's machine, avoiding the errors and inconsistencies that could otherwise creep into the process at the printer's end.[22] These companies help buyers in fragmented markets choose products by providing up-to-the-minute price information and a single contact point for service.

All of these changes will bring savings in indirect purchases. With direct purchases, matters are more complicated. Most companies buy custom-built parts and components from suppliers with which they hold long-term contracts and work on design specifications. Rogue purchases are not a problem: no local manager nips out to buy a few thousand motherboards. The answer here is not a standard catalogue. Instead, Internet-based software enables companies both to widen their potential pool of suppliers for any given contract and to deepen their relationship with the supplier that eventually wins it.

In general, companies more often use the Internet to deepen their relationship with existing suppliers than to broaden it with new ones. Many companies are beginning to put contracts for supplies out to tender on-line. That can bring substantial savings. GE Lighting claims to have brought down its average cost of supply by 20 percent by asking for quotations by e-mail from a wider range of suppliers.[23] But inviting tenders on-line is not simple. It requires more standard processes as well as elaborate preparation and, before a company asks for bids, it must first ensure that every specification apart from price is nailed down. The sheer complexity of that process often makes companies reluctant to switch to suppliers that may not understand their needs.

The impetus is to work with suppliers in new and more sophisticated ways, rather than merely to order the cheapest product from the lowest-cost producer.[24] As with consumers on-line, companies seem more interested in using the Internet to specify with greater precision what they want than in saving money. But, by periodically testing the market, companies may enjoy both benefits. As a paper[25] by Sam Kinney, one of the founders of FreeMarkets, a Pittsburgh company that runs

. . . . .

electronic auctions, puts it: "Buyers typically use the auction to determine with whom to establish the market relationship, based on excellent price discovery. But, once the auction is over, production parts are approved, and tooling is installed, the working relationship can run for years." That "price discovery" is not to be sniffed at. FreeMarkets claims that, over a five-year period, buyers have ended up paying on average 15 to 16 percent less than the previous purchase price.

Auctions

Companies have a vast array of electronic markets and on-line auctions to choose from. For consumers, the sheer breadth of an on-line auction market such as eBay or Europe's QXL makes it a novelty in a world dominated by high-street stores and fixed prices. But businesses are more used to dealing in markets and more accustomed to flexible prices. Thus every kind of market developed for consumers is also springing up in business-to-business commerce.

The most venerable business markets have sold commodities: money, metals, meat. Financial markets moved from trading floors to wires and computers in the 1980s. Commodity markets have taken longer, but many commodities, from steel to rubber, are now being traded on-line. In addition, the Internet's low start-up costs and broad reach, coupled with the computer's ability to handle complexity, have encouraged the advent of newer sorts of marketplace to deal in products that are often characterized by their perishability. Telephone bandwidth is one example; several companies are competing to run an electronic trading platform for the mass of small companies that buy spare minutes from the larger telephone companies. Energy is another. In the United States, only 2 percent of natural-gas trading and 0.2 percent of electricity trading was conducted on-line in 1999; but in April 2000 six giant energy companies set up an on-line trading consortium. The biggest on-line trader in the world is said to be Enron, which has an international trading operation based in London (the city with the world's richest supply of foreign-language speakers).[26]

Another novel commodity auctioned on-line is advertising space. OneMediaPlace holds Web-based auctions to sell spare advertising capacity on Web sites, radio and broadcasting channels, billboards, and newspapers. For many advertising sales staff, leftover space is a nightmare. It has to be sold cheaply and quickly, sometimes to customers that have already bought space at a higher price. The idea behind One-MediaPlace is to allow bargain-hunting media buyers to place their bids without having to negotiate laboriously with sales reps. They can automate their bids, too, by entering the maximum price they are prepared to pay for a certain kind of space.[27]

Some of the new on-line models do something rather simpler: they match bids and offers, collecting payment, enforcing rules, and taking a turn on the way. Ace-Quote, a British company, allows buyers to post, free of charge, requests for computer-related products or services. Sellers pay an annual fee to receive the requests and place bids.[28] Another example is the National Transportation Exchange (NTE), which uses the Internet to connect shippers looking for bargain prices with fleet managers that have spare space in their trucks. It sets daily prices based on information from several hundred fleet managers about the destinations of their vehicles and the amount of space available. The NTE issues the contract, handles the payment, and collects a commission. The fleet manager gets extra revenue; the shipper, a bargain price. In time, individual truckers will be able to connect directly with the market via wireless.[29]

People and skills can be traded as well as space and time. Smarterwork, an on-line service launched in May 2000, is an electronic exchange for buyers and sellers of any work that can be fulfilled electronically. Jobs carried out by suppliers are divided into five categories: document production, graphic design, writing and editing, Web-site work, and Internet-based research. The company employs in-house experts who check the credentials of all those who offer their services on the site, provides detailed profiles of buyers and sellers, and manages payments.[30]

Some of these markets are intended mainly to provide a common trading standard, so that buyers and sellers can easily exchange information electronically. An example is Covisint, a joint venture between four of the world's biggest car companies. The aim of such projects is to assemble an entire industry, not into a supply chain but into a network – or, to use the fashionable term, an eco-system. For that, common standards are an essential first step.

Some of these markets have attracted the attention of regulators in the United States and Brussels. They worry that a market with only a small number of buyers or sellers is more prone to capture than a market in which many deal with many.

It seems probable that the number of markets will shrink rapidly, if only because markets work better the more buyers and sellers they have. Once single national exchanges exist, they will begin to coalesce across borders, creating global markets for goods and (more important) services that have previously been traded nationally.

It seems unlikely, however, that the very openness of the Internet should restrict competition rather than widening it. If a market turns out to be heavily biased, toward buyers or sellers, a grey market is likely to emerge alongside it, draining off business. Only in industries that are already too concentrated for competition to thrive might on-line exchanges make the problem worse.

Integrating suppliers

Most of the time, companies are not testing and re-testing their contracts with their suppliers at on-line auctions. But the Internet changes the relationship in other ways. It enables companies to outsource many activities they would once have owned and run directly. And it enables them to share information with their suppliers much faster and more fully than would previously have been the case.

One of the best examples is that of Cisco Systems. Every day, the company posts its requirements for components on an extranet, a dedicated Internet-based network that connects the company to thirty-two manufacturing plants. Cisco does not own any of these plants, but they have been through a lengthy process of certification to ensure that they meet the company's quality-control standards. Within hours, these suppliers respond with a price, a delivery time, and a record of their recent performance on reliability and product quality.

Another example is Dell Computer. It allows its suppliers real-time access to orders on its extranet, making it possible for suppliers to organize their own production and delivery schedules to fill demand at exactly the right moment. Dell, along with a growing number of imitators, uses this streamlined supply chain to enable customers to specify just what product they want.

This integration of the supply chain has other advantages for business-to-business transactions, because buyers of products such as machine tools or other intermediate goods are likely to want their purchases designed to their own specifications. Weyerhauser, a forest-products company, uses an extranet to enable customers to specify the exact features of a door, say, then feeds that information directly into the manufacturing process. The effect has been to improve on-time delivery, market share, and return on net assets.[31]

Improvements in information can push changes right through the value chain. An example of that process at work comes from ChemStation, an American manufacturer of industrial detergents, which realized even before the Internet emerged that its business of shipping detergents needed to be reorganized. Because finished detergents consisted largely of water, shipping costs frequently exceeded the cost of the product itself. The company therefore set up a number of small reconstitution plants. With a stock of the basic ingredients and computerized recipes to mix them on the spot, ChemStation can now make exactly what its customers want.

The change has also affected the way ChemStation's customers operate. One steel mill used to run an entire operation to unload deliveries of detergent, decant it into drums, store the drums, and deliver them around the plant – and then reversed the whole process. Now Chem-Station has placed a hundred or so tanks where they are needed around the mill, and uses electronic monitoring to ensure the tanks remain full and that the customer uses the right amount of the active ingredient.[32]

Suppliers benefit greatly when they can see their customers' production schedules and sales data. They can react at once rather than waiting for news to trickle down. Dell, which, like Cisco, has remarkably few suppliers, is working on ways to pass on the information that it now shares with its own suppliers to the second tier of companies that supply them in turn. It dreams of the day when all computer manufacturers that buy hard drives, and all suppliers that produce them, make that information available anonymously on an electronic exchange, allowing the whole industry a clear view of the balance of supply and demand.[33] For older companies, integration of this kind is difficult. Ford, for example, would dearly love to emulate Dell. But it faces many more obstacles. Dell is a relatively young company that sells directly; Ford has a large chain of suppliers, many of which see the Internet as a threat to their livelihood. Dell has designed its assembly process around direct sales and has great flexibility to alter the design of a product to meet consumer demand. Ford not only has a much more complex product line; it manufactures in a way appropriate to its different distribution system. For instance, the colour of a car is stamped into the metal at an early stage, making it difficult to offer consumers last-minute choice.

The gains from integration are tremendous, however. One of the most important is to reduce inventory and improve cash flow. Old-style companies borrow money from the bank, use it to pay for the manufacture of a product and its storage until a customer comes along, and then sell the product. Companies that integrate supply chains do things the other way round: first, they take an order and payment from the customer, then they build the product. The result is lower inventory and less need for working capital.

"Inventory...is the physical embodiment of bad information," says Paul Bell, who runs Dell's operations in Europe.[34] In industries where prices fall fast, inventory is also potentially lethal. The average price of components, which account for 80 percent of the cost of a computer, drops by 30 percent a year. So every day that can be cut from stocks saves a computer manufacturer money. Dell claims that in its case parts sit in inventory for only eight days before they are shipped to customers.[35] In addition, the company effectively uses its customers' credit-card companies, rather than its own bankers, to finance production.

Integration instantly passes information about customer decisions right along the supply chain. If customers suddenly stop buying green cars, the word can be fed immediately from on-line purchasing systems and catalogues to every supplier involved. The effect is to reduce inventory, especially of finished goods.

That, indeed, had already begun to happen before the Internet came along, thanks to a combination of EDI and enterprise resource planning, which allows better integration of procurement and production. Overall, American durable-goods manufacturers reduced inventories as a share of sales by more than a quarter between 1989 and 1999 – and that takes no account of savings in the costs associated with financing stock, warehousing it, and discounting it to meet shifts in demand.[36]

The effect of the Internet will be to make these savings, so far confined largely to big companies such as car manufacturers, available to smaller firms and to companies such as clothing manufacturers that have more fragmented and geographically scattered supply chains. In the long run, the result may be to reduce the ferocity of recessions. In the past, when growth in demand has slowed down, inventories have often fallen, amplifying a mild deceleration into a fullblown recession.

Employees

Over time, the Internet will alter the relationship between companies and employees. It becomes easier for employees to compare their pay and hunt for other jobs, and for companies to buy in services for particular projects. The relationship will therefore often become looser and more job specific. At the same time, other employees, recruited for their experience, inventiveness, or managerial skills, will bring with them what economists call "tacit capital": the knowledge that becomes a company's main underlying value. These employees, whose prototypes work for so many Silicon Valley start-ups, will be rewarded for their knowledge mainly by becoming part-owners of the companies they join. This is capitalism's solution to the Marxist dilemma: the workers end up controlling the capital.

Hiring and paying

Just as markets for goods and services are becoming broader and more transparent as they move on-line, so is the market for human labor. One market shifting rapidly on-line is that for corporate recruitment. Most companies have a "come-and-join-us" page on their Web site which is likely to attract more visitors than any other page.[37] Advertising a job on an e-recruitment site such as Monsterboard, which claims to have four million resumés and 400,000 jobs on offer, or StepStone, a Norwegian equivalent, typically costs about 5 percent of what it would cost to place an ad in a big newspaper, and the reach can be at least as great.[38] Some companies have begun to abandon entirely the advertising of jobs in print, especially if they are hiring graduates, scientists, or IT staff.

The process of recruitment should now be better informed, on both sides, and swifter. Every prudent graduate going for a job interview these days begins with a trip to that corporate Web site. More and more companies insist that all replies arrive on-line. Recruitment companies try to design forms that will filter out unsuitable candidates automatically. A few are also trying to develop psychometric and other tests that can be used on-line to cut down the time-consuming selection procedure.

Once hired, employees have more ways to discover how well they are paid. By clicking on an on-line salary guide such as iwantanewjob.com, a worker can check roughly what he or she might earn as a database administrator in an American university, say, or as a software developer in the airline business.

Most people's job prospects will long continue to be determined mainly by local factors. But, helped partly by the Internet, a world market is emerging for certain skills. More highly skilled workers move across borders than ever before. Even when they cannot do so physically, they may still find work abroad.

America's communications revolution has been driven partly by skilled workers pouring in from abroad. Immigrants, two-thirds of them from Asia, account for almost one-third of Silicon Valley's work force. Chinese and Indian engineers between them started 29 percent of Silicon Valley's technology companies between 1995 and 1998. A quarter of Microsoft's employees are foreign-born.[39]

Thwarted by immigration policies, though, many more workers stay at home and export their skills. The United States takes about half of all software exports from India, for example, many of them in the form of outsourced and customized work rather than finished products. Both India and Israel are home to software-development centers for Hewlett-Packard, IBM, Intel, and Microsoft.[40] And Novosoft, an American Web software company, employs Siberian programmers who charge $45 an hour, compared with the $100 to $320 an hour charged by programmers in the United States or Britain.[41]

Such international outsourcing will lead to new inequalities in pay. The gap between what people are paid within countries will widen. The gap between what people earn in different parts of the world for similar work will narrow. A glimpse of what may lie ahead is to be found in Estonia. In that small ex-communist country, dock workers earn only a quarter as much as dock workers in Finland, just across the Baltic. But skilled software programmers in Estonia earn almost as much as their Finnish counterparts – and more, given Finland's hefty taxes. As a result, some young Finns with software skills now work from Tallinn, Estonia's capital, and sell their skills on-line.[42]

Immigration officials may still bar the door to foreigners who want to apply in the flesh, but more and more services can be delivered on-line. Where people with scarce skills can live in one country and work in another, they can command an international rate. For such people, the Internet offers a global job-offers page.

The new team

As workers with rare skills acquire a better sense of their market value, the balance of power between them and their employers changes. They become just one more category of supplier – although a particularly important and diverse one.

Specialist workers are likely to become ever more footloose. They may work independently, bidding for particular projects as freelancers. Or they may work within a company, but move from project to project in teams that bill the company for their services rather than their skills. The Internet will make both versions possible. Freelancers have an inexpensive way to search for work, and corporate intranets can improve the management of internal teams.

At the same time, companies are developing new ways to manage teams of employees. In the case of design projects, the Internet enables scattered employees to collaborate on plans. Car companies have long used private networks to share development work on complex designs among technicians scattered around the world. Now, a wider range of people can share in a project – marketing folk, for example, or corporate-communications staff. The result is a change in the way product development takes place: more people, in more parts of a business, can take part simultaneously. An example of this in action is provided by Telenor, a Norwegian telecoms company, which has created "virtual meeting places" – parts of the company's intranet where staff working on new projects can file everything from product design to minutes of meetings. The aim is to create a common place to keep all material, rather than scattering it around filing cabinets. Staff find that they still need frequent face-to-face meetings, but the company claims it has halved the time-to-market for some projects. Among other benefits has been the ability to pull in people from different offices to work on the same project, because they no longer need to move in order to cooperate.[43]

Selling knowledge

Although companies will buy in more and more of the skills they need – not just catering and construction, but legal, accounting, and financial skills – the core value of many businesses will consist of the brains of a handful of key employees. Because of their symbiotic link with their companies, these people will be treated more as part-owners of the business than as staff.

They will thus be paid at least partly in stock options, which give employees a share in the risks and rewards of an innovative new company. One American survey found that 92 percent of firms backed by venture capitalists, many of them in Internet-related businesses, gave their employees stock options.[44] Options palled as the prices of high-tech stocks dwindled, but they remain the simplest way to align the interests of employees and owners.

For established companies, the task of acquiring and retaining bright people is harder. True, they offer a security that start-ups cannot match. But prospects for their share price may be unexciting, and they cannot offer the thrill and profit of a hot flotation if they are already on the market. Moreover, their shareholders may dislike new share options because they dilute existing holdings. A solution is to create tracker stocks or incubators to simulate the rewards (but usually not the risks) of a new venture.

Bright people, moving from company to company, may be a nightmare for firms that find it hard to keep staff. But they serve a useful function as far as the Internet is concerned, for they take with them the knowledge gleaned in each previous job, speeding up the application of new ideas and the spread of knowledge. Not every new thought can be disseminated on-line; thus itinerant communities of knowledge workers are the pollinators of the new crop of ventures and business processes.

The Future of the Firm

· · · · ·

As they integrate the Internet into their operations, companies will change. They will combine services with manufacturing in new ways. They will become more focused on the customer and less on the process of production. And they will form more alliances, at many different levels. Above all, they will often become looser confederations of teams and suppliers, glued together by corporate culture and rich communications. As the Internet reworks the relationship between companies and their suppliers and customers, the company effectively becomes an intermediary, although of a rather special kind.

Facing the customer

Once companies begin to exploit the Internet fully, they find themselves almost unintentionally swept around to face their customers and to respond to their demands. Suddenly, they have the tools to spot which customers are profitable – and, sometimes, to predict when those valuable clients are likely to leave. At last, they are able to do something about a problem they have long discussed: the fact that it invariably costs more to recruit a new customer than to sell to an existing one.

Add to that understanding the realization that the customer now drives the whole production process. It is the click of a mouse on Dell's computer site that starts the first nuts and bolts rolling down the production line, not an order from the production manager. Management structures, designed to focus on the production process, look out of date. What matters now is to concentrate on the customer.

That task becomes more rewarding once companies realize that the customer can become a partner in development, as well as in production. An example is Agility.com, a software start-up firm in Boston,

Massachusetts. It used the Web to survey customers' views of a new software product called Agile Manager. It set up a Web site where potential customers could see slides of the product and use a "nonfunctional" demonstration. It then e-mailed the address to selected people and asked to arrange a time for a short interview about the product. The company found that this technique gave it several advantages. It could interview people in their homes and offices, saving them time. It could talk inexpensively to people overseas. And it got more information than a questionnaire would have yielded.[45]

Product development will increasingly take such interactive forms. Jungle.com, a British on-line entertainment retailer, tracks the pattern of customer inquiries. If it spots that customers are confused, it alters the information on its Web site. Blue Square, a British betting site, uses the information it gleans from customer service to tailor its next promotions more precisely.[46] Companies are thus looking for ways to build customers into a community that will tell them when a product is wrong and teach them how to put it right. This is a new approach, but it will be one of the main hallmarks of the company of the future.

Outsourcing

Thanks to the Internet, companies find it easier to outsource activities they would once have carried out in-house. A classic example of this trend is Nortel Networks, a Canadian company that specializes in building high-performance communications networks. In 1998 the company decided to sell many of its plants to large contract manufacturers that make parts for networks. The purchasing firms signed long-term supply agreements with Nortel.

The benefits to Nortel from this move away from vertical integration arise partly from the fact that the contract manufacturers have larger turnovers. They can thus afford to invest more in developing specialized components. They can also reap economies of scale. Nortel itself gains flexibility: if it has a large order from a particular part of the world, it can more easily arrange production near by. It has also gained the freedom to concentrate on what it does best: making highly sophisticated and valuable parts and installing communications networks. The Internet enables it to keep in constant touch with its suppliers at many different levels.

Because the Internet makes it easier for companies to manage outsourcing, and because almost everything that a company needs, from management skills to the human-resources department, can now be bought in, it is possible to create a company from nothing in no time. As a result, competition springs up more quickly. A large number of Internet start-ups are "plug-and-play" companies.[47]

Many of the barriers to entry that once protected big companies are therefore disappearing. Naturally, the new competitors home in mainly on the older company's most profitable activities. GE, at the behest of Jack Welch, its chief executive, ran an exercise in 1999 called "destroy-yourbusiness.com": managers were told to go off and work out where they were most vulnerable to competition delivered by the Internet. The threat of such competition creates an extra incentive to avoid too much vertical integration and to concentrate instead on what a company does best, outsourcing the rest.

Alliances

The Internet encourages alliances. Again, they were already proliferating before it began to worm its way into corporate structure. In 1998, a quarter of the earnings of the top one thousand firms in the United States came from strategic alliances, double the share in the early 1990s.[48]

The number of alliances has been growing both between the world's main regions and within them, and the growth has been especially noticeable in areas of rapid technical advance such as information technology.[49] As in so many other ways, the Internet here reinforces existing trends, working with the grain of corporate change rather than against it. That is why its influence is so great.

The Internet gives companies new incentives to work together: to build a trading hub, say, or to offer links to one another's products on their Web sites. Alliances provide an insurance policy against risk – the risk of new technology and the risk of new business models. They also allow companies to deepen their relationship with customers by offering them bundles of products, some created by alliance partners. On top of all these incentives, the Internet provides the means. It enables the construction of links not just between the folk at the top of two allied companies, but at many levels on the way down. Those links can be thick or thin, rich or sparse. But whichever they are, they require companies to trust one another more than they have usually done in the past, and to be willing to be open.

The Hollywood company

Will the communications revolution make the company an anachronism? The question is not a new one. In *The Nature of the Firm*, published in 1937, Ronald Coase, an economist who was later awarded a Nobel prize for his work, asked why workers were organized in firms instead of acting as independent buyers and sellers of goods at each stage of production. He concluded that buying and selling entailed many significant costs associated with preparing and monitoring agreements with suppliers and customers, and that the existence of companies minimized these transaction costs. It takes time or money or both to find out what products are being bought and sold. Companies hold down those costs.

A newspaper, for example, may own its printing presses or it may contract with outside printers. In Britain, the *Mirror* and the *Telegraph* groups own their presses, the *Financial Times* and the *Independent* do not. Printing capacity may be less expensive if bought under contract, but some of the savings will be offset by the higher costs of coordination.

As communications improve, they reduce such transaction costs. Scouting the market for spare capacity and low prices becomes easier. Technologies such as on-line catalogues and electronic billing reduce the costs of dealing with customers and suppliers. All kinds of service can be outsourced. The result will be companies organized into smaller units.

Companies will come to look more like Hollywood studios. Between the two world wars, Hollywood stars tended to be employed on contract by studios. That had advantages: it stabilized the stars' incomes and guaranteed them work, but it also meant that the stars effectively subsidized the less good or less popular performers. In time, the studio system broke up. Making a movie today involves assembling a temporary "company": buying in the services of script-writers, costume-designers, technicians, producers, Leonardo di Caprio, and full supporting cast. Each participant earns an income that reflects his or her individual worth in the market, rather than a pay scale imposed from above on a group of people of disparate talent.

Many companies will come to resemble a movie in the making, assembling groups of people to work on particular projects. For knowl-edge-based companies, this fragmentation comes naturally. They can farm out a larger share of activities than can a firm in a traditional heavy industry. Computers and communications therefore make it pos-sible for companies to become networks of independent workers, spe-cializing in what they do best and buying in everything else. Vertical integration will diminish; horizontal integration, often achieved by alliances and partnerships, will flourish.

These companies of the future will need skills that today's managers rarely cultivate, at least in Anglo-Saxon countries: the ability to collab-orate, to develop unique skills while at the same time operating with a high degree of openness, a capacity for trust, a facility for involving cus-tomers in a continuous dialogue. So the Internet will create, not just new markets and new business structures, new supply chains and new customer relationships, but new kinds of corporate manager as well.

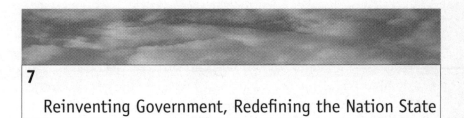

Reinventing Government, Redefining the Nation State

Moving house in Singapore? Look no further than eCitizen, the comprehensive Web portal of the government of Singapore. Among the many points on the walk through life with which the government offers help ("pursue a first-class career," "employ people"), is a package compiled by the Housing and Development Board, the government agency that builds and administers the estates on which most of Singapore's people live. Through this, you can submit an on-line application for a flat, then go on to complete electronic forms for your parking permit, television license, telephone, and utilities such as water and electricity. A few clicks send notification of a change of address to every government agency that needs it, including the postal service, so that it can redirect mail.[1]

The eCitizen project, launched in 1999 and described by America's General Services Administration as "the most developed example of integrated service delivery in the world," captures several key points about the potential the communications revolution holds for government.

There is, to begin with, huge opportunity. The main impacts of the communications revolution will be on service industries, and the world's biggest service industry by far is government. The business side of government – procurement and the delivery of services – will eventually benefit from the cost savings and increases in efficiency that so many companies are already starting to enjoy.

The political pressure to exploit the power of communications to improve public services will grow as the new technology works its magic on the quality of private services. As these become increasingly focused on the customer, available round the clock reliably and conveniently, the inadequacies of public provision will grow more apparent.

Even bigger gains for citizens may come from rethinking the whole process of government, removing the complexity that confronts people in the offline world and indeed changing the structure of government, just as the structure of many companies is changing as they absorb the impact of the Internet. In the words of Stephen Goldsmith, a former mayor of Indianapolis, e-government is a "just-in-time innovation to correct the brittleness of twentieth century command-and-control government."[2] Citizens, in their dealings with government departments, will acquire more control over their lives, just as customers, in their dealings with companies, have been empowered. A survey conducted by Forrester Research in 2000 shows how this is just starting to happen. [Fig 7-1]

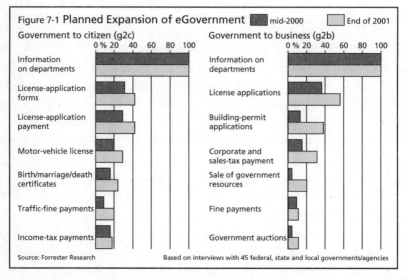

But it is not only the business side of government that will change. The communications revolution will also have pervasive effects on government's size and role. The powers governments have over their citizens will alter. A stronger link will be forged between what voters want their government to do and what they are prepared to pay in taxation. As commercial activity becomes more mobile and less a prisoner of geography, some taxes will become almost voluntary. The consequent reduction in revenue will force governments to reconsider which services they provide, and how they provide them. They will target services more precisely, cut costs, and concentrate on regulating and monitoring. Government will therefore become smaller.

In time, the revolution will transform not only the administrative machinery of government, but the balance of power between governments and their citizens. People can become better informed, even though most governments are no more enthusiastic about putting information into cyberspace than they were about publishing it in more traditional ways. People can also communicate their views on their rulers more easily, even though governments may still often take what them with a pinch of salt. Indeed, the electoral process itself may change, although the first experiment with digital democracy, in Arizona's Democratic primary in March 2000, hit a host of hitches. It becomes possible to give people more power to participate directly in government decisions, if that is what voters want. In the United States, one result may be to increase even more the influence of lobbying and public-opinion polls.

What will this mean for the power of the nation state? When, in 1948, George Orwell envisaged the world of 1984, he imagined global dictatorship based on electronic communications that could monitor citizens' every action. Now, when electronic networks can connect databases and video cameras around the world to unprecedented computer power, Orwell's vision has become a practical proposition. Governments are acquiring the technological capability to isolate and track every movement of those citizens they regard with suspicion. This surveillance power will be a serious potential danger to liberty, especially as countries without Western respect for human rights acquire Western networks and computers.

On the other hand, the world's nastiest governments have tended to have bad communications and a majority of citizens have been too poor to leave data trails. Is that an accident? More probably, there is a link between good communications and political freedom. If so, the death of distance will foster democracy.

It will also undermine the authority of the nation state, shifting power in subtle ways. It will force even reluctant governments to do more things jointly with the private sector: outsourcing, just as companies increasingly outsource. More activities may have to be self-policing: governments' jurisdictions are geographic, but the Internet transcends geography. And governments will have to cooperate more with each other and learn to work through international bodies. Otherwise, they will find national rules and standards frequently undermined wherever they differ from those elsewhere.

The communications revolution may even affect the desirable size of the nation state. In the past, the size of political units has often been determined by geography (although, as Eli Noam of Columbia University has pointed out, in Napoleonic France the sizes of *départements*, the main administrative units, were established to permit an official to travel to any part of them and still be home in time for dinner).[3] The death of distance will reduce the handicaps that isolation has previously imposed on countries on the fringes of economic regions. The dissolution of some geographical disadvantages will be especially important to the many small countries that came into existence in the second half of the twentieth century.

Finally, the death of distance will, on balance, be a force for peace. Because it favors democracy, and democracies are more reluctant to fight than dictatorships, it will reduce the potential for conflict. Moreover, the best way to discourage countries from fighting one another is surely through better communication. Now, ordinary citizens everywhere will become more familiar with the ideas and aspirations of people around the world. The citizens of one country may come to understand those in other countries a little better, and the glue that binds humanity will be strengthened.

The Political Process

Al Gore, in his early days as vice president of the United States, talked often of using communications technology for "forging a new Athenian age of democracy." (Rather antiquated technology was eventually Mr Gore's downfall in the 2000 presidential election.) That might not be such a good idea. After all, Athenian democracy excluded women and slaves – a majority of the population – from the rights of citizenship, and one of the main concerns for government in applying communications technology must be to avoid disadvantaging citizens on the wrong side of the "digital divide." But Gore's essential point remains valid.

A change will clearly occur in the relationship between politicians and the governed – almost certainly for the better.

That is not surprising. Over the past century, electronic communications have already been the biggest single force for altering the way democracies work, allowing people with ideas to rally voters, and voters to be heard by politicians. Radio and television have revolutionized both the way political debate takes place and the things politicians must do to get elected and re-elected.

Now, free communications will shift the balance of power between government and the citizen in several new, important ways. Citizens can, if they choose, become better informed; they can more easily make their views known; and the process of voting itself may alter, especially if people want to vote on individual policy decisions in a sort of continuous referendum.

Informing the citizen

Good information is essential for effective political involvement, and the communications revolution makes information more accessible than ever before. It also allows people a wider choice of sources of information, especially important in repressive countries where the national news media are biased or controlled by government.

Information that was once expensive to obtain can now be located inexpensively. That is important. Until the Internet came along, a great deal of supposedly public information was actually available only to an elite: the media, officials, big business. Now, it has become easier for citizens in many countries to discover facts that affect their lives. For instance, a local group of American environmentalists can now consult the Toxic Release Inventory (in which chemical companies based in the United States must list some of the nasty emissions that their plants give off), search through it by zip code, and pick out the relevant figures for the local factory. And during South Korea's parliamentary elections in 2000, a grassroots citizens group posted on the Internet information that the conventional media were too timid to run, about candidates who had criminal records or were draft-dodgers or tax-evaders. Of the eighty-six candidates on their blacklist, fifty-eight lost, including several political heavyweights. Only thanks to the Internet was the group itself, made up of six hundred smaller groups, able to cooperate to build its protest movement. The site registered 1.1 million hits on election day.[4]

Use of communications technology (hacking excepted) will not help citizens to root out facts that a government is determined to keep from them. And, although many more people may be able to read it, information from, say, the European Commission will not necessarily be more useful – or more comprehensible – simple because it is available electronically. Gradually, though, the need to make information accessible may help to change the culture of bureaucracies. As long as the only audience for a document is other officials, they may tolerate obfuscation in a way that the public at large will not.

Above all, people will have access to a diversity of information, and this will help to keep governments honest. Government propaganda becomes less convincing once people hear the other sides in an argument. Up until now, the main source of alternative news for ordinary folk in many countries has been radio – mainly the BBC World Service. Transmitting BBC programs requires public subsidy and is often at least partly dependent on the goodwill of the local government. The overseas service, now available on the Internet, finds its audience beginning to expand in new regions. It may be that wider and inexpensive access to global sources of information will help to act as an antidote to prejudice, nationalism, and war-mongering.

A voice for citizens

Whether they want to or not, governments will become more aware of what their citizens think of their policies. The Internet and other communications improvements reduce the costs of campaigning and lobbying, and ease the task of finding out which politicians claim to believe what.

Those who want to take to the streets find on the Internet inexpensive ways to drum up a march or even a riot. International campaigns, such as those against the World Trade Organisation and genetically modified crops, now have the perfect mechanism to link arms around the world. For isolated campaigners, such as Russia's beleaguered environmentalists, this is wonderful. For multinationals, and multinational bodies, it is a nightmare.

Lobbying politicians is universal in democracies, but electronic communications make the task quicker and simpler. One American site with the nifty name of E The People allows users to run their own online petition on a vast range of subjects, from the right to scavenge in dumpsters to the finer points of Vermont's car-licensing laws. E-mail also enables enraged voters to send their views to politicians on impulse. At the click of a mouse, a lobbyist can fire off a message to every member of Congress. A letter or fax, by comparison, takes time and preparation. Bill Clinton, as president, received up to two thousand items of e-mail a day, which made him by far the world's most e-mailed person. (Of course, when one gets that much e-mail, it is difficult to read everything. In 1994 Carl Bildt, then prime minister of Sweden and an inveterate nethead, sent an e-mail to President Clinton. It was a historic moment: the first Internet exchange between two such senior politicians. What happened? Nothing. After two days of waiting, Mr Bildt's staff rang the White House instead.)[5]

Certainly it is becoming easier for voters to discover which politicians are on their side. Before the American presidential election of 2000, sites such as CandidateCompare and SelectSmart enabled voters to type in their prejudices and find a politician who shared them. Desirable though it may be that individuals can communicate more readily with politicians, some people fear that the main effect has been to increase the productivity of lobbyists. As a result, politicians are influenced by the hubbub of electronically delivered special pleading and fail to think about the interests of the country as a whole.

More probably, politicians will discriminate between the types of message they receive. Some American politicians complain that 80 percent of their e-mail comes from people who do not live in the state or district they represent. The main danger of excessive lobbying is that it devalues electronic messages. The old-fashioned letter on paper, typed – or, better, laboriously hand-written – has ironically become more influential than the urgent screed on the screen.

. . . . .

Direct democracy

In the final two decades of the twentieth century, democracy became the standard form of government around the world. During that time, at least thirty-two formerly authoritarian regimes held relatively free elections.[6] But voters often feel dissatisfied and defect from political parties to lobbying groups. Turnouts flag: only 50.7 percent of American voters took part in the 1996 presidential election. Such revolts against the democratic process, say some analysts, reflect a feeling among voters that their voices go unheard. The average citizen votes only every few years, usually with an infinitesimal influence on the outcome.

Now, several initiatives aim to use the Internet to revive the political process in various ways. In his ultimately unsuccessful bid for the Republican presidential nomination in 2000, Senator John McCain made startling use of the Internet to raise campaign finance and coordinate grassroots volunteers. In the first forty-eight hours after his triumph in the Republican primary in New Hampshire, he raised $1 million in campaign funding through his official Web site. At least one commercial Web site, PoliticsOn-line, sells Internet tools and software to candidates and campaign managers who want to build a fund-raising site.[7] The 2000 presidential campaign also saw a large experiment with electronic voter registration. More than one million voters registered by June and it was predicted that five million would have done so by November.

Beyond these developments may lie another one: on-line voting, although the experience of the Arizona primary was not wholly encouraging. The voting site went down for an hour on the first day of voting and the helpline was overwhelmed. Users of Macintosh computers ran into particular difficulties, prompting Mark Fleischer, the local party chairman, to observe, "They're the wrong group to make mad, I'll tell you that."[8] However, 40,000 people voted on-line, representing a 600 percent increase in turnout over 1996. Many local and regional bodies will experiment with on-line voting in the years ahead.

Some visionaries see a future in which "televoting" becomes the norm, and even allows the development of what enthusiasts such as Alvin and Heidi Toffler, two of the gurus of cyberpolling, call "semi-direct democracy." In their book, *Creating a New Civilization*,[9] the Tofflers argue that voters should be allowed to make many more policy decisions. After all, if people can shop from home, no obvious technical reason prevents them from eventually voting from home.

The supporters of direct democracy point to various advantages that electronic voting from home might bring. They argue, for example, that it might boost flagging election turnouts. In 2000, 104.5 million Americans voted in the presidential election; 132.5 million watched the Super Bowl. If it were as convenient to vote as to watch the Super Bowl, more people might do so. They argue, too, that some form of televoting might help to define constituencies in new ways. Already, in the United States, the drawing of political boundaries attempts to ensure an appropriate racial mix; in future, "virtual" constituencies could be created, comprising voters randomly selected from all over a state or a country, or constituencies could be designed to represent different ethnic groups or special interests.

Take such views to their logical conclusion, and it might be possible to eliminate elected representatives entirely. Home after a day's work, people might be offered a slate of propositions on which to vote. Proponents of such a view imagine a government by the entire citizenry, electronically assembled. Such a system would reduce the power of lobbyists. It might be easy to persuade a few hundred representatives to a particular viewpoint, but it would be far harder to persuade an entire electorate.

Such claims for direct democracy interest social and political theorists in particular because of the rise in referendums, a traditional form of direct democracy. Since the start of the 1970s, the frequency of referendums has almost doubled.[10] Many have been held in the United States, where referendums have an unusual characteristic: they are held mainly in response to voters' propositions or initiatives. A lobbying group with the requisite number of supporters can, in many states, put a proposition to the electorate at large. The Internet makes it easier to muster support for such a proposition.

But gathering support is one thing; allowing voters to take binding decisions with the click of a mouse is quite another. Referendums may indeed provide a fertile testing ground for electronic voting. But there are huge problems to overcome first. On-line polls will need up-to-date on-line databases, for example, as well as a way to authenticate voters and a powerful and hacker-proof system of encryption. Even with those technical barriers out of the way, many will worry that, as long as Internet access is dominated by the better-off, on-line voting will discriminate against the poor and ethnic minorities. Maybe; although that would depend on how the system were administered. Those groups tend to have the lowest turnouts at the moment. Anything that made them more likely to vote would be an advance.

As for the visionary dream of Athenian democracy, with armchair voting every night, it is likely to remain a dream. Voters almost certainly prefer to delegate most of the business of running the country to specialists. The political intermediary – the politician – will remain, performing a specialist role on behalf of voters reluctant to carry the burden of deciding everything from the size of the state budget to the appropriate weight limit for trucks.

However, even if voters do not want to make political decisions directly, representation will increasingly become a convenience, rather than a necessity. Thanks to advances in communications and information technology, the business of informing voters and seeking their views will become more sophisticated. So will the business of lobbying them. And the more political power shifts from politicians to people, the more interest groups will have to lobby the people, rather than their representatives.

A Smaller, More Efficient State

Just as the communications revolution alters the relationship between voters and government, so it facilitates the business of government. The role of the state will increasingly be one of informing, monitoring, and measuring services rather than providing them directly. Just as companies will become more fluid organizations, employing fewer people, so government will increasingly become a coordinator rather than a service provider.

But the transformation of government will take longer than the change in companies is doing. Although the Internet began life as a government project, and governments still own a large stake in the telecommunications systems of many countries, most governments have seen their primary role as exhorting the private sector to find new uses for communications. They have been slow to see the opportunities on their own doorstep. Even public procurement, vast by the standards of all but the largest private companies, has been slow to move on-line. But as it does, its catalytic impact on the private sector will be huge. If governments want to speed up the spread of e-commerce in their countries, the best way to do so would be to move as much government activity on-line as possible.

So far, governments and their agencies have been followers, not leaders. There are many reasons for this tardiness, some reasonable, some less so. One of the most important is that state agencies are not under the same pressures as private companies. They are generally monopolies, and so the Internet does not threaten their core business. Moreover, they rarely care much about customer satisfaction: people do not deal usually with them by choice, but because they have to.

Certainly, they rarely feel obliged to go in search of customers. A survey by Forrester Research, an American consultancy, of forty-five American government agencies with on-line initiatives found that two-thirds of them did not actively market their sites, relying instead on word of mouth to promote them. Several spoke of the cultural inhibitions that discouraged marketing and emphasis on customer service. "There exists a command-and-control mentality in government that is inflexible," said one respondent from a Federal agency. "This mindset is inconsistent with the approach we need to successfully implement digital government."[11] Plenty of other, more practical factors hindered change. [Fig 7-2]

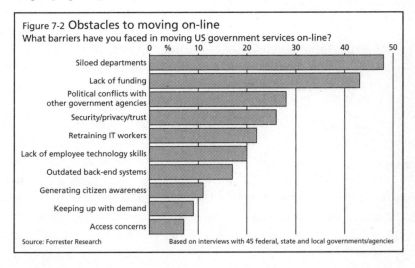

Figure 7-2 **Obstacles to moving on-line**
What barriers have you faced in moving US government services on-line?

Siloed departments
Lack of funding
Political conflicts with other government agencies
Security/privacy/trust
Retraining IT workers
Lack of employee technology skills
Outdated back-end systems
Generating citizen awareness
Keeping up with demand
Access concerns

Source: Forrester Research   Based on interviews with 45 federal, state and local governments/agencies

· · · · ·

That inflexibility is one of the greatest obstacles to change. The structure of government is vastly more complex than that of any private company, and even more protected from change by venerable tradition and internal politics.

As governments try to respond, they will find out what private firms have already discovered. The communications revolution increases the power (and capability) of customers, and thus makes them, and not the production process, the focus of organizational structure. Just as companies are being forced to dismantle the vertically structured "silos" into which different departments have so long organized production, so public services will be forced to adopt what in Britain is called "joined-up government." But there will be plenty of battles along the way. At least twenty federal agencies provide educational loans in the United States, for example. Once government tries to create a single loan "portal," any number of bureaucratic turf wars will break out as the education and commerce departments skirmish for control of funding and management.[12]

There are also more defensible reasons for the slow pace at which government agencies have found uses for communications. First, most of the people they serve tend to be poor, uneducated, or elderly – and therefore among the least likely to be on-line. And it is not only the disadvantaged who may lack access. Even in the United States, just over 40 percent of households yet have an Internet connection at home, and in most of Europe the figure is far smaller.

Second, governments will have just as many problems as companies with rebuilding "legacy" computer systems to serve the requirements of customers, rather than those of an individual government agency. The need to link up with private companies will aggravate such problems. Sometimes, governments may have to create entirely new databases.

Third, as governments start to do more, privacy will become an issue. Governments everywhere know far more about their citizens than does any private company: their blood group, their income, their tangles with the law, their vehicle registration number. Some agencies, such as America's Department of Motor Vehicles, even sell such data. Thus security, and trust, will become even more important issues for government than they are for private companies. People have arguably handed over so much information to government agencies precisely because they assumed that government was incapable of collating it – let alone commercially exploiting it. As governments acquire the technical power to do so, citizens may grow more reluctant to part with personal details.

Given these many obstacles, what will push governments to move? Two factors will be important. First, the communications revolution will erode tax revenues. Although a motor manufacturer can in theory shut down an assembly plant and move production elsewhere, that is disruptive. But finance, software, and telecommunications businesses are lighter footed. Their capital value is largely in the heads of their staff, and they will therefore locate in the places where those people want to live.

The erosion of revenue will probably be greatest for indirect taxes; but direct taxes and corporate taxes will be affected as well. In response, governments will hunt for ways to reduce state spending. Electronic communications and information technology make it easier to contract out some services, target other services more precisely, and raise the productivity of those services for which the state continues to pay. The state will thus employ fewer people and spend less revenue, shifting its role from provider to promoter and regulator.

The second factor will be the pace of change in the private sector. The quality of private services has been improving at an unprecedented pace as competition sharpens and customers' expectations rise. Customers expect convenience, choice, swift delivery, round-the-clock service, and personalization from private firms. Asked by Forrester Research what they saw as the main benefit of dealing with government electronically, a group of five thousand respondents put at the top of the list the freedom to transact outside government office hours.[13] Why, they will increasingly demand, should public services be less customer-friendly than private ones?

Moreover, as private companies (and individuals) move on-line, the need to deal with government on paper will grow more onerous. The relative cost of transactions with the state will increase, unless government changes too.

· · · · ·

State spending

Electronic communications will offer new ways to deliver services, from vehicle registration to medical care. Initially, they will be used for simple tasks such as form-filling. In time, though, procurement will move on-line, generating large savings in cost. Other savings will come from outsourcing as many services as possible, which new communications will again make easier. Among the eventual advantages will be greater reliability and accountability.

Form-filling, fine-paying

One of the fastest ways for the public sector to raise productivity and at the same time improve quality of service is by moving on-line many of the routine tasks that citizens daily queue to undertake in state and municipal offices. Just as companies get their customers to take on some of the form-filling chores they once paid staff to provide, so government agencies can get citizens to perform more of the work themselves.

At first, such sites often simply provide the opportunity to ask questions or download a form, which then has to be filled in and mailed back. America's Internal Revenue Service was one of the earliest government bodies to offer such a service, and from the start it was hugely popular. The site had almost three million hits on the day before forms were due to be returned in 1997 – at the time, probably a record for a single day[14].

Beyond this simple use of the Internet is the rather more sophisticated possibility of a two-way trade of information. North Dakota now accepts fishing-license applications on-line, for example, while the State of Arizona's transportation department runs ServiceArizona, which enables people to renew a car or truck registration or to replace a lost driver's license or ID card. The service is available twenty-four hours a day – unlike the municipal office that still offers the services to unconnected folk – and users love it. But it also saves money: processing an on-line request costs only $1.60, compared with $6.60 for the same transaction over the counter. Because ServiceArizona, which has been running since 1996, now processes 15 percent of renewals, the motor-vehicle department already saves $1.7 million a year, some of which goes to improve the off-line service.[15] Other Web sites, mostly provided by state and local governments, allow users to fill in and return a form to pay a traffic fine or enrol in a college course. In Chile, the Internal Taxation Service allows tax returns to be submitted entirely on-line. Citizens can check their returns and review their past tax history. The site has led to a sharp reduction both in inaccuracies and in the time taken to process tax returns.[16]

Sites such as these can guide users through the intricacies of form-filling, picking up mistakes as they are made and noting inconsistencies. As a result, the form that finally arrives on the bureaucrat's desk is already more accurate and complete than would previously have been the case. The sites also offer a way to simplify and streamline the tedious exchanges of detailed information that make up so much of the process of administering taxation and benefit entitlements that every state undertakes. Some sites go further. The Pennsylvania Department of Labor and Industry has a site that allows people who have been denied benefit after being fired to lodge an appeal. The site establishes the circumstances of dismissal by asking the claimant a series of questions, thus reducing the scope for legal argument. That site might easily be a template for a whole range of relatively routine cases on which, at present, government lawyers laboriously sift evidence and give judgments.

But governments are often slow adopters. Not surprisingly, especially in the United States, nimbler private companies are popping up to offer to outsource some of the tasks that governments have yet to move on-line. Some concentrate on one function. PayTheTicket.com aims to allow people in the United States to pay traffic fines with credit cards on-line, while Accela.com issues building permits through a service called "Velocity Hall." (The on-line entrepreneurs offer a wittier sort of service than the standard bureaucratic product.) Others are more ambitious. GovWorks claims to process credit-card payments of fines, permits, utility bills, and local taxes for 3,600 American municipalities. It also offers services such as on-line auctions of government surplus and a who's who of state and county officials. But because GovWorks is not linked directly into most of the governments with which it does business, its processing is slow. The big gains, in speed, service, and savings, will come only as government learns to emulate the on-line sophistication of private industries such as financial services.

The governments at the head of the pack are already moving on from form-filling to creating on-line one-stop shops: central Internet portals, like Singapore's eCitizen, which invite people to log on with a password and find the function – not the department or agency – they need to deal with a particular problem. Such portals are being developed by communities such as the Danish town of Naestved, which allows citizens who have applied for a digital signature to look at their own files and perform legal transactions over the Internet.[17] NaestvedNet is part of a single Internet database aimed at providing access to all regional services – for government employees, private companies, and citizens – from a single site.

Apart from Singapore – which, with three million people, is smaller than Toronto – few governments are yet building such sites (although the state of Victoria in Australia has one called MAXI, through which people can buy a liquor license, apply to take a driving test, or register to vote).[18] As they develop, such sites should transform the arduous business of dealing with government. You have a spell of ill-health, you change your job, you want to retire early: such portals will offer a one-stop shop for the paperwork or questions that might once have forced you to trek from one bureaucracy to another.

A bigger transformation, though, will occur within government, which will at last start to think about provision from the customer's standpoint, rather than from the minister's or bureaucrat's. Governments will begin to wonder whether many tasks would be better handled by the private sector. Do state employees need to collect taxes, if private companies can handle fine payments? Does a government agency need to distribute permits, if privately run portals are shepherding people to its Web site? The lines between the public and private sector will blur, as government moves from providing services directly to monitoring their provision by others.

Procurement

For private companies, many of the earliest savings of time and money have come from moving the supply chain on-line. Government can potentially enjoy all of the same advantages: tenders from a wider circle of suppliers, scale economies through coordinating purchasing by different departments, the benefits of speed, and the reduction of inventory that can result. Simply to list these benefits is to hint at possible problems, however. Government procurement, unlike corporate purchasing, is a morass of special rules and provisions that suppliers must first meet. But even these rules may be more easily enforced if companies have to apply to tender on-line. The potential savings are vast, given the sheer scale of government purchasing. Moreover, just as the shift to on-line buying by large companies pushes their suppliers in the same direction, so on-line procurement by government is probably the biggest single tool that governments possess to accelerate the move to on-line tendering and supply right across the economy.

Governments are already starting to shift purchases on-line. In the United States in 1998, the Defense Logistics Agency constructed a site called Emall to handle relatively small purchases (under $2,500). The procurement officer can place orders on-line and receive them the next day. The cost of some purchases in 1999 was halved as a result.[19] Australia's Victoria state has had similar results from shifting much of the purchasing carried out by its Department of Natural Resources and Environment on-line. Not only has purchasing efficiency improved by 70 percent, but fraud has fallen and compliance with purchasing rules has improved.

Whereas routine administration may be the area in which governments can make the largest financial savings, there will be many other services where provision also improves. Among them are welfare payments, crime prevention, health care, and education.

Welfare payments

One of the snags with using communications technology to improve the administration and payment of welfare benefits is that the poor and elderly, the main recipients, are the group least likely to have the right sort of access. Yet communications and database technology, combined with smart cards (pieces of plastic that store information on a microchip) can still be used to increase efficiency and reduce fraud.

In the United States, an electronic transfer system for paying welfare benefits was installed in the state of Texas by a company called Transactive on behalf of the Texas Department of Human Services. Instead of food stamps, welfare claimants receive a Lone Star Card. When they enter a PIN number and swipe the card through a terminal at the check-out counter in any of fifteen thousand retail outlets, the cost of their food purchases is deducted automatically from their entitlement and added to the retailer's account. The system recognizes the barcodes of forbidden items such as alcohol and cigarettes, thus ensuring that food stamps are actually spent on food. Texas officials noticed that, when towns introduced the new procedure, sales of alcohol tended to fall and sales of food to rise. By weeding out fraudulent claims, the system has produced substantial savings.[20]

An even more imaginative payments system has been introduced in South Africa, where the payment of state retirement pensions in remote parts of the bush is being transformed by the combination of satellite technology and touch-screen automated teller machines (ATMs). At present, most state pensions are paid by check. In villages where there are no banks, checks must often be cashed at a local shop, which means the pensioner pays a commission. Many pensioners, moreover, are illiterate, making it difficult for them to know if they are being cheated. Conversely, the provincial governments paying the pensions lose money through fraudulent claims.

To tackle these problems, First National, South Africa's biggest bank, has devised a procedure, subsidized by provincial governments, whereby pensioners receive a bank card containing an ordinary magnetic strip along with an electronic record of their index-finger print. To receive their money, the pensioners insert their bank cards into a cash dispenser and put their index fingers against an electronic scanner. An ATM delivers the cash only if the print matches the one on the card. To serve rural areas, the bank straps a mobile ATM to a truck that visits different villages on a fixed day each month – pursued by an entourage of migrant traders offering everything from chickens to cloth. The system has also been adopted in neighboring Namibia.[21]

Crime prevention

Inexpensive computing power and communications can transform public safety. The use of video cameras, for example, has already reduced crime in trouble spots. When sixty remote-controlled video cameras were installed in the English town of King's Lynn, crime fell almost immediately to one-seventieth of its previous level. The savings in patrol costs alone rapidly paid for the equipment. In other parts of the United Kingdom, more than 250,000 cameras have been installed in areas of high crime.[22] A study of the effects of cameras in Newcastle found that arrest rates also rose after they were installed, by a quarter for burglary, criminal damage, and offences of drunkenness. Increasingly, individuals in high-crime areas may install their own security cameras, allowing self-service policing to develop.

By using the Internet's potential to search for and deliver information, police and private companies are also fighting crime. Police forces publish information on wanted criminals on the Internet. After riots in London in June 1999, the City of London police posted more than a hundred video shots on their Web site, resulting in more than fifty identifications.[23] More controversial are private sites such as CrimeNet, an on-line database of criminal records based in Melbourne, Australia, which offers information on more than four thousand crimes, lists confidence tricksters, records lost and stolen property, and carries appeals relating to unsolved crimes. The site allows users who make a credit-card payment of A$6 to search for an individual's criminal record. The aim is to make such information available to people who want to check the credentials of a potential business partner, nanny, or neighbor.[24]

Such sites, like many uses of new communications technology to fight crime, worry civil-rights activists. In 2000 an Australian judge aborted a murder trial because details of previous crimes by the accused, of which the jury was supposed to be unaware, were posted on CrimeNet. The site responded by adding a terms-of-use contract, asking users first to declare that they were not serving on a jury.[25] Self-policing may be the only way to deal with conflicts of this sort. Shutting down such a site in Australia would not prevent the information from appearing on-line elsewhere. In 1995, Reuters reported some details of the preliminary hearings in the case of Rosemary West (subsequently convicted of a notorious series of murders), on its non-British news feed.

. . . . .

The details, which could not legally be reported in Britain, were posted on a Web site hosted outside Britain. Under pressure from British police, Reuters eventually stopped the coverage. But police and courts cannot always rely on such voluntary compliance.

Health care

After welfare, health care is the largest single item on the budgets of most rich countries. Even in the United States, publicly financed health care absorbs a larger share of GDP than defence. The burden will rise with the ageing of the population, forcing governments to look for imaginative ways to reduce costs. The easiest single way to do this is to keep patients out of hospital and to use the time of trained staff in productive ways.

Keeping patients out of hospital means treating them more in their own homes. In the United States, several projects are developing telemedicine, to try to reduce the five hundred million home visits that the country's nurses and health aides make each year. Nurses at the Hays Medical Center in Kansas, for example, are using telephones and computers to monitor elderly patients who live in the vast spaces of the western Kansas plains. The system enables a nurse to check up to fifteen patients an hour, while a nurse physically traveling from house to house may see only five or six a day. Each televisit costs $35, compared with $90 for a home visit by a nurse. In addition, home telemedicine seems to reduce visits to hospital emergency departments and to allow patients who may eventually need extended care to spend longer in their own homes.[26] Elsewhere, telemedicine, backed up by home visits, is now being used to monitor patients with diabetes, haemophilia, skin disorders, and a whole range of other ailments. Monitoring devices will check whether patients have taken their medication each day and automatically contact a nurse if they fail to give the right answer (failure to take medicine at the right times is a leading cause of hospital admissions).[27] A result of such innovations will be to increase the extent to which health care can be delivered in the home, rather than in the doctor's surgery or hospital.

An even simpler system turns out to be Britain's NHS Direct, a telephone and Internet helpline launched in 1998. Staffed by nurses, and available when doctors' surgeries are shut, it costs £8 ($12) a call, compared with £10 if a patient consults a general practitioner and £42 for an admission to a hospital emergency room. A study of a pilot version of the service found that it reduced both out-of-hours calls to doctors and emergency admissions to hospital.[28]

Eventually it may be possible to save on hospital costs by carrying out operations far from large hospitals, using the skills of doctors many miles away to perform remote surgery. That would also allow small hospitals to save money by drawing on the skills of a distant pool of surgeons. The techniques are being developed by the Pentagon, its interest driven by the need to find better ways to operate on wounded soldiers in battlefields where doctors cannot safely go. In past wars, up to 90 percent of the deaths of those who were injured in frontline battle occurred because the wounded did not reach hospital in time. The American Army Medical Department now aims to bring a mobile operating room to the wounded. A surgeon, far away but watching on a three-dimensional monitor, would operate using a central master control that would manipulate remote forceps, scalpel, and needles. What may become possible for a wounded soldier will also, in time, be available for civilians.[29]

Education

The expense of higher education is soaring. Even in 1995, private universities in the United States effectively charged each student nearly $60 a lecture, a figure that will have increased since and did not take account of either public or private support. Yet, with rapid changes in the job market, the demand for adult and continuing education is expanding sharply. It is rising even faster in the developing world. Could the answer to all these demands be more distance learning, which (unlike conventional learning) can offer economies of scale and a quality-controlled product to a vast virtual campus?

Distance learning is not new. Britain's Open University has been offering courses over radio and television networks for almost thirty years, while the oldest distance-learning university of all is the University of South Africa, at which both Nelson Mandela and Zimbabwe's Robert Mugabe earned their degrees. Where money is short or students

. . . . .

are widely scattered, distance learning may be the only alternative to no higher education for many people. This is certainly true for the biggest such project of all: the Central China Television University, with somewhere between one million and two million students – more than the rest of the world's distance learners put together.

Now, some argue that distance learning will challenge the sway of conventional universities. Peter Drucker, doyen of American management gurus, gives them only thirty more years of life. Others, more convincingly, see distance learning as a way to overcome barriers of distance not only in terms of geography, but of social class. John Seely Brown and Paul Duguid, two authors who (rightly) rate the survival prospects of conventional universities more highly, note that students who are reluctant to speak in face-to-face classes – including non-native speakers, women, and the disabled – seem to participate more actively when the interaction is on-line.[30] But the best results in distance education seem to require a mixture of on-line delivery and face-to-face discussion. In the 1999 – 2000 academic year, the University of Glasgow delivered one course of lectures in the conventional way to its local students and, simultaneously, over a video link to a new campus it has helped to establish at Crichton College, near Dumfries in southwest Scotland. After they had listened to the lecture, students at Glasgow dispersed; those at Crichton discussed it with local tutors. In the end-of-term exams, the Crichton students outperformed the very similar group who had followed the course on the spot in Glasgow.[31]

As private companies start to enter this field, they find that the most profitable and successful areas are delivering training courses to companies. That may change. One survey of consumers found that only about 31 percent felt that college tuition gave "good" or even "average" value for money.[32] Distance learning may not have the cachet of a good university name, but it will be as good as, and less expensive than, a mediocre one.

Exporting services

Once public services can be provided remotely, they can also be exported, generating revenue, or imported, cutting costs. If America's traffic fines can be processed in Atlanta, why not in Bangalore?

Already, the death of distance is making it possible for some countries to provide for one another services that have traditionally been in the public sector. For the moment, these are generally arrangements of convenience, not commerce. For example, under a program at the Massachusetts General Hospital, a team of seventy radiologists has X-rays wired from its own telemedicine centre in Riyadh, Saudi Arabia.[33] The University of Maryland's University College, which specializes in part-time education, teaches courses at American military bases throughout Europe and Asia, laying on computer conferences and voice-mail to allow students to communicate with their teachers and with one another.

In time, governments may procure their back-office processing from the best and least expensive provider, just as banks and insurance companies are starting to do. Sometimes, the provider may be in a different country and speak a different language. As that happens, governments will come to realize – as financial-services companies are realizing – that their role is to deal directly with the public, not to manufacture or process paperwork, a task in which the state has no obvious comparative advantage. Once governments are no longer giant paper-shuffling machines, they will be able to focus more clearly on the roles that only they can fill, such as regulation, standard-setting, and legislating. Governments that achieve such clarity will be smaller, more efficient and – the greatest prize of all – more popular with their voters than any government is today.

Redefining the Nation State

· · · · ·

Government jurisdictions are geographic. The Internet knows few boundaries. The clash between the two will reduce what individual countries can do. Government sovereignty, already eroded by forces such as trade liberalization, will diminish further. As an incisive study of the issues puts it, the Internet is "redefining the powers and behavior of governments...Increasingly, the jurisdiction of government authority is different from the economic marketplace."[34]

One result: no longer will governments be able to set the tax rates or other standards they want. Another: international cooperation among governments, and between government and the private sector, will become more important. One more: the advantages that size once conferred on countries will diminish. As borders break down, some of the handicaps that small countries have suffered will diminish.

Taxation

For governments, the most disturbing single impact of new communications may be the erosion of the tax base. Several factors will be at work: the greater geographic mobility of companies that trade in information, rather than physical products; the greater anonymity that the Internet can offer; and the sheer difficulty, as companies fragment, of defining exactly where or how economic value is being created. All of these problems are aggravated in federal countries, such as the United States, where taxes may vary between regions as well as between the country and its neighbors. As there is no easy answer, governments will have to rely more on voluntary compliance and less on their own muscle.

Taxes on sales, companies, and people will all be harder to collect. A fundamental requirement for most taxation, especially on income and sales, is knowledge of location: the place where a person is resident, the national origin of the income a company receives, the place where a transaction takes place. Now, not only will people and companies be able to undertake many more transactions at a distance – by, for instance, selling services in one country that are bought in another; it will also be harder to decide where a particular transaction has taken place.

Indirect Taxes.

Indirect taxes have already created difficulties. In the United States, electronic retailing owes its popularity partly to the fact that it is a gigantic tax scam. Austan Goolsbee, an economist at the University of Chicago, estimates that on-line sales in 1998 would have been up to 24 percent lower if existing sales taxes had been charged on them.[35]

Why does electronic commerce generally escape sales tax? For the moment, the courts have ruled that an out-of-state company has no duty to collect a sales tax on goods coming into the state unless that company has a physical presence, or "nexus," in that state. Buy a book in a store in Manhattan, and you will pay 8.25 percent in state and city sales taxes. Buy the same book on-line from Amazon.com and you will pay nothing. Indeed, one of the reasons Jeff Bezos, Amazon's founder, chose Seattle as a location was because it had a small population and so few potential sales-tax payers.[36]

This issue will grow more complex as more high-street retailers start to sell on-line. If a high-street firm collects the tax on its on-line sales, it will be undercut by its Internet-only rivals. But if it fails to collect, its ample physical "nexuses" put it at greater risk of prosecution than on-line firms usually are. Partly to try to get around this problem, Barnes & Noble set up its on-line operations as a separate legal entity.[37]

Such conundrums survive because America's 1998 Internet Tax Freedom Act legislated for a three-year ban on "new Internet taxes." When that moratorium expires in 2001, will "old" sales taxes be levied in the new economy – and if so, how? An advisory body (the Gilmore Commission) set up to find an answer could not, when it finished its deliberations in March 2000, come up with an agreed reply. But it recommended (among other things) that digital products downloaded over the Internet, such as software, books, and music, should be tax-exempt – and so, in the interests of tax neutrality, should their physical equivalents.

Tricky as it is to find a policy on indirect taxes in the United States, it will be even harder where cross-border trade is concerned. The future of indirect taxes worries many governments and government agencies. In the United States, sales tax accounts for roughly 12 percent of state and local-government revenues, although in some places, such as Texas, it is much more. In the European Union, value-added tax (VAT) typically accounts for about 30 percent of total tax revenues. In developing countries, indirect tax can be even more important.[38]

When businesses can sell goods and services across borders by mail order or on-line, which state or country taxes them? Taxing goods shipped across borders or state lines is tough; taxing intangibles is well nigh impossible. If governments try too hard to impose indirect taxes on physical goods ordered on-line, some, such as CDs and videos, may mutate into intangibles and escape taxation.

The world's two largest trading blocks have fundamentally different approaches to indirect taxation. America's sales taxes tend to fall only on goods, not services, and to be levied on the final customer. All Euro-

. . . . .

pean countries levy VAT, a more complicated tax designed half a century ago, when most products were physical and consumed in the country in which they were produced. VAT is levied nationally. Exporters receive a tax rebate; importers are assessed for tax. Services tend to carry higher rates of tax than goods. So the rate of VAT paid on a piece of recorded music will differ depending on whether it is sold as a CD or downloaded from a site (and so classified as a service).

If the site is in another European country, the EU says that retailers should collect the tax at the rate that applies in the member state where the goods will be consumed. But customs officials in, say, Italy are unlikely to try to ensure that their local exporters hand over uncollected tax revenue on their sales to Germany to the German government. If the site is in the United States, things are even murkier. The EU has boldly proposed that foreign firms should decide which European country to locate in for tax purposes, and then levy the appropriate rate of VAT and send it to the appropriate national government. That proposal is undoubtedly in the flying-pigs department.

Direct taxes.

Companies and individuals will gain new tax freedom too. Taxing companies grows more difficult as it becomes harder to see exactly where value is produced. Physical presence, one of the key tests to decide where a company should pay tax, becomes less important as a company's assets increasingly take the form of software code and good ideas. The structure of companies becomes more fragmented, making it harder to capture exactly where the production process happens. And information-based firms are simply more mobile than companies producing large volumes of physical products.

Such companies will increasingly choose their location mainly on the basis of where their best people want to work. As Swedish companies, operating in one of the rich world's most punitive income-tax regimes, know full well, tax can be a determining factor.

Many high earners will have more freedom to live where they choose. In 1999 the Italian government took Luciano Pavarotti to court, claiming that he had been a resident in Modena and therefore owed more than $2 million in taxes. Mr Pavarotti retorted that he lived in Monte Carlo and spent only a few days in Italy during the years in question. He eventually settled with a six-figure sum. The court case, say some, was a challenge by the Italian government to the rapidly growing number of its citizens who have moved abroad to avoid taxes.

Pavarottis will always be rare: only a small proportion of citizens are ever likely to move in order to avoid tax. But those who do will be people who pay a disproportionate share of taxes and generate a disproportionate share of jobs. In addition, businesses that employ large numbers of highly paid professionals may migrate to low-tax countries. As a result, countries that up to now have tended to compete for inward investment will increasingly find themselves competing in a global market for citizens, trying to attract inward professional and management talent. The winners will be those countries that can offer the combination of the lowest taxes and the highest quality of life.

Overall, the impact of new communications is likely to be to reduce tax revenues. One effect may be the growth of new forms of tax (or rather, extremely old ones). Property was the earliest tax base – because it is hard to hide – and may also be the newest. But governments will have to learn to make do with a smaller share of their electorate's cash, or work more closely with other countries to devise harmonized tax policies.

New international cooperation

In many other areas, governments will find that their jurisdictions overlap or that their authority is eroded. They will face a choice: to fight fruitlessly to protect their diminishing sovereignty, or to find ways to manage their relations with other countries and the private sector so that they get at least some of what they want.

One answer is to work through international organizations, a growing number of which are tackling points of cross-border governance. Some are traditional bodies, such as the World Trade Organization (WTO) and the World Intellectual Property Organization (WIPO), which are seeking to adapt existing policies to the electronic world. The WTO, for example, is the forum in which ministers agreed in 1998 a temporary moratorium on customs duties for products delivered on-line, while the WIPO has been trying to apply the principles of intellectual-property ownership to domain names. Others are new bodies, some of them the partnerships that coordinate parts of the Internet,

• • • • •

such as the World Wide Web Consortium (W3C) and the Internet Corporation for Assigned Names and Numbers (ICANN).

Some of these groups, argue Catherine Mann and her colleagues in a study for the Institute of International Economics, have found that they can develop international approaches, especially to problems of finding a legal framework and developing a trusted environment in which electronic commerce can flourish. Other issues, such as privacy and the protection of personal data, are harder to deal with in an international context, partly because social attitudes and government approaches to policymaking are different. In the case of data protection, for instance, America's instincts have been to leave the task to the private sector and self-regulation; Europe's instinct has been to pass laws. This divide will run through many debates on the Internet's future between these two blocks.

An alternative approach to the erosion of jurisdiction is to leave more to the private sector. Several business groups have emerged, such as the Global Business Dialogue on Electronic Commerce, to try to make sure that the views of companies are represented in government discussions and to try to develop self-regulation. In several areas, governments hope that the private sector can play a bigger role – credit-card companies, for example, might become the main agents of consumer protection, because they are used to dealing with complaints from shoppers. In practice, a combination of government regulation and self-regulation will emerge.

Neither course necessarily means global harmonization of standards. The European Union, for example, whose members have already accepted a large sacrifice of sovereignty to achieve what Catherine Mann and her co-authors call "interoperable" economies, has evolved a combination of harmonization and mutual recognition. Thus a bank that meets the regulatory requirements of, say, France has the automatic right to establishment in, say, Germany. Many countries would probably prefer to recognize the standards of others than to have their own domestic standards dictated by other countries.

Ultimately, what governments do in this area will be determined largely by what consumers want. The Internet's global reach creates new opportunities for competition in regulation and standard-setting. In countries that acquire a reputation for trustworthy law-enforcement and vigorous consumer protection, businesses will gain. They will become trust havens. Regulatory competition may seem an odd basis for building international agreement, but it will be what counts.

Smaller countries

As government shrinks, so too may the size of nations. Indeed, that was the trend throughout the second half of the twentieth century. The number of independent countries almost doubled in the forty years after 1960, thanks mainly to the end of colonial rule. In the late 1980s and early 1990s, the collapse of the Soviet Union, the division of Czechoslovakia, and the disintegration of Yugoslavia all added to the number of nation states. In the twenty-first century more nations may emerge, some peacefully, some violently, if separatist movements continue to gather strength.

Many of the new countries are small. Indeed, half of the world's countries now have fewer people than the state of Massachusetts has, with six million. The death of distance will reduce the handicap such niche nations face.

In the past, the main economic advantage of being part of a large country was to have access to a large marketplace. Small countries thus have a particular interest in anything that enables them to treat the entire world as their home market. Well run small countries, such as Hong Kong, Singapore, and Luxembourg, have economies that are wide open to international trade. They therefore have a greater interest than larger countries in a liberal international trading regime that allows them to export freely without the hindrance of tariff barriers.

The need to be part of a global market gives small countries a special need for inexpensive communications. They often occupy the fringes of larger markets. Like small companies, they need to become global niche players, selling specialist products to the biggest market of all: the world market. The Internet is therefore a boon to them. It allows them to advertise their wares worldwide. Countries such as Ireland (population: 3.6 million) and Iceland (population: 260,000) are using their well educated work forces to foster communications-based businesses. Ireland is especially remarkable. It has built its higher-education system around information technology: about 70 percent of its college students study engineering, science, computer science, or business. As a result, it has become one of Europe's biggest call-center and help-desk operators and attracted foreign investment from most of the big names in American computers.

Low-cost communications will not, of course, remove all the economic handicaps that small countries suffer. Their greater specialization still makes them more vulnerable to external shocks than are bigger nations. And telecommunications need to be supported with other inexpensive communications, such as good international air links. But for independence movements, the communications revolution is good news. Not only can they fight their wars of liberation with the laptop and a Web site, but, once they succeed, their new minination has a better chance of survival than it would have done in the past. The combination of a liberal global trading regime and good communications would allow an independent Scotland or Quebec to compete on less unequal terms with the world's bigger countries.

Communications and Peace

· · · · ·

Inscribed over the entrance to Bush House, for many years the home of the BBC World Service in London, is the motto of the corporation, founded in Britain between the first and second world wars: "Nation shall speak peace unto nation." Might the death of distance also cement the quarrelsome countries of the world together in new ways, reducing conflict and spreading peace? Pessimists point to plenty of examples where communications have been used for precisely the opposite purpose: to Radio Milles Collines which stirred Rwanda's Hutus to murder their Tutsi neighbors, or to the way press coverage of Somalia, Bosnia, and Kosovo lured the United States into conflicts in which it had no obvious strategic interest.

Undoubtedly, too, the growing dependence of countries on communications and information technology makes them vulnerable to new threats: to jamming, viruses, and interference. Defense strategy is now as much about communications and computer skills as about building better bombs. The tools of the hacker have become as important as the (more romantic) skills of the spy.

However, there are also reasons to hope that better communications will indeed reduce conflicts and wars, and not just because they avoid the sort of misunderstandings that led the British and Americans, in 1815, to fight the battle of New Orleans unnecessarily because news of the end of the war of 1812 in Europe did not reach North America until a fortnight later.[39] Democracies have always been less inclined to fight than other forms of government; and countries with good communications tend to be predominantly democratic.

Good communications between governments are the bedrock of peace. Governments are more likely to fall out if they lack a clear idea of one another's intentions. In the new millennium, governments can be better informed than ever about what other governments are doing. Diplomats can find on the Internet material that would once have been extracted only by a skilled ambassador.

In addition, the death of distance means that countries will be tied together by innumerable commercial bonds. Many more companies will have subsidiaries and operations on every continent, or at least in all three main time zones. Countries that invest in one another are also less likely to fight.

Above all, ordinary citizens will learn more about people in other countries. Few individuals use mail order or the telephone to shop abroad; even fewer watch foreign television channels or read foreign newspapers. Such activities will now be easier and less expensive. People will in time grow used to the idea of shopping abroad, not just for goods but also for services. Buy a foreign car, and you learn nothing about the country where it was made. Buy a foreign movie or a CD or use a foreign secretarial service, and you start to build a relationship of sorts with another nation.

As international trade in services builds up, it will create a firmer bond between countries than trade in goods has done. Such activities will help to shrink the world and make people realize the extent to which, in John Donne's words, "No man is an island, entire of itself; every man is a piece of the continent, a part of the main."

Globalization and revolutionary communications have blossomed before, bringing peace and prosperity. The years before the first world war were also a time when nations grew rich on the proceeds of free trade, ordered by telegraph, carried by rail and steam. Many enthusiasts thought that the telegraph would foster peace. Yet communications did not prevent the rich world from destroying its gains by embarking on a pointless conflict. This time, perhaps, it will be different. Free to explore different points of view, on the Internet or on the thousands of television and radio channels that will eventually be available, people may become less susceptible to propaganda from politicians who seek to stir up conflicts. Bonded together by the invisible strands of global communications, humanity may find that lasting peace and prosperity are underpinned by the death of distance.

8

A New Economy

Two tremendous changes are driving through the world economy: technological advance in computing and communications, and the fall in barriers to trade and investment. The death of distance brings the two together. As the world moves toward virtually limitless and almost free electronic communications capacity, trade and investment flows will transform patterns of economic activity.

These changes have had paradoxical effects. On the one hand, they have fed the longest continuous economic expansion in the world's largest economy. By the start of the new millennium, the United States had expanded for an unprecedented decade, growing faster than the European Union in all but three years of the final two decades of the old century, and faster than Japan in all but three years of the final decade. The boom was accompanied, again mainly in the United States, by a soaring bull market in equities – especially high-technology stocks – and by the astonishing growth of small dotcom businesses. Many new technologies breed bull markets, but the Internet beat the field. In 1995 the sole public Internet company, America Online, was worth $1 billion; in June 2000, there were more than four hundred public Internet companies, capitalized at more than $750 billion.[1] The bubble had by then already begun to deflate, but it leaves behind a high level of innovation and investment in information technology. No less remarkable is the way that inflation, which has choked off so many previous booms, remained low by recent historic standards.

Many people, watching this exuberance, wonder whether the basic rules of economics have changed for good. They talk of "a new economy" or a "new paradigm." Modern economies, they argue, are less

. . . . .

vulnerable to cyclical fluctuations than the industrial economies of the past, and so are more stable. They can be run at lower levels of unemployment without disagreeable outbreaks of inflation. And productivity growth will be higher, leading some to predict a "long boom" of the sort that the world last enjoyed at the turn of the nineteenth century, another period of breakneck technological progress.

At the same time, many also fear the consequences of economic change. They see technological advance as a threat to jobs. They fear the development of a disenfranchised underclass, deprived of job security and access to the complex technologies on which societies will increasingly depend. They fear, too, the consequences of globalization, and see trade and foreign investment as further threats. They worry that the industrializing economies, with their immense supply of low-paid workers, will deprive higher-paid workers in rich countries of their jobs. They already see money moving as never before: on average, some $1.3 trillion, or about one-third of a year's global exports, flashes around the world every day. How, in the face of such numbers, can countries control their economic destinies? Put technology and globalization together, and you have a recipe for even greater terrors. In a world in which communication costs almost nothing, any service can be sold on-line anywhere in the world. Jobs will move to the people. But what if the people live on the other side of the world?

How should one strike the balance between the optimists of the new paradigm and the doomsayers of job destruction? The optimists are right that there is likely to be a permanent shift in the long-term productivity of the economy, but wrong to assume the end of booms and slumps. The pessimists are right that the changes will have repercussions for jobs and for income distribution that will be traumatic for many people, but fail to notice their enriching effects.

The most fundamental impact of the communications revolution on the economy will come from the way it transforms the availability of information and knowledge and the cost of transferring ideas. All sorts of information – a business plan, a request for tenders, a blueprint – are now infinitely diffusible. The barriers to the instant global spread of knowledge are falling away. Anyone's bright idea can quickly become everyone's bright idea.

This has two immense consequences. First, knowledge is the basic building block of economic growth. It can also improve lives. A new technique for curing a sick child, for tilling a field, or for designing a car will travel rapidly around the world, spreading health, employment, and wealth with it. It is important to remember that economic growth has always been built on technological advance. From the printing press to the steam engine to electricity, innovations have made society richer, not poorer. The communications revolution will do the same.

Second, the benefits of extra wealth will, on the whole, accrue to consumers, in the form of better quality and lower prices, rather than to companies, in the form of higher profits. The diffusion of knowledge and information makes markets work better. It makes them more transparent, allowing new competition to appear. Although some new monopolies will spring up, especially if governments protect them with patents (see Chapter 10), others will be toppled. In the new economy, there is less of an imbalance of information and market power between the knowing seller and the ignorant buyer. Some economists have dubbed it the "nude economy," because transactions of all kinds become more see-through.

With these larger economic trends will come others in the structure of the economy. The rules of location are being altered. Some industries are suddenly free from the shackles of geography while others need more than ever to be in clusters of related businesses. The old divide between goods and services is giving way to a new divide, between products requiring physical delivery and products that can be delivered on-line. Moreover, for many products, and especially the "weightless" sort, global markets will become more important than local ones. The change will not happen instantly. National regulatory barriers now hinder international trade in services far more than they do trade in goods. But, ultimately, trade and foreign investment will become even more important, and economies even more interdependent, than they are today.

Such changes will undoubtedly disrupt the lives of many individuals. They, along with companies and governments, are having to adjust more rapidly than they have ever done before. Yet the rich countries already have older populations than ever, and they will be older yet by mid-century, when half the people of the rich world will be more than forty-six years old. So societies less demographically fitted for change than at any time in the past face faster and more pervasive change than at any previous time. It is the social dimension of profound economic and technological change that will be hardest for many countries to come to terms with.

Paradigm or Paradox?

· · · · ·

As with previous great innovations, one of the most widely expected effects of the communications revolution is on productivity, and thus on the underlying rate of economic growth. Growth can be generated by increases in labor or in capital, or by increases in the productivity of both as a result of using them more efficiently. In the two centuries since the industrial revolution, the driving force behind the global rise in wealth has been greater productivity, first in agriculture and then in manufacturing. New mechanical and electrical technologies, together with the development of the factory and mass-production techniques, have enabled people to produce vastly more per head than ever before.

The revolution in information technology driving the death of distance seems likely to have a similarly dramatic impact on productivity in the twenty-first century, and particularly in service industries. Measuring that impact has proved difficult, however. As Robert Solow, an American economist and Nobel-prize winner, put it in 1987, "We see the computer age everywhere except in the productivity statistics."[2] The phenomenon Professor Solow described has become known as the "productivity paradox." While there are already signs that information technology, communications, and the Internet are transforming individual industries, slashing costs and increasing efficiency, it is harder to prove the effect on productivity. Yet in time, the communications revolution might prove to be an engine of global economic growth just as powerful as steam, railways, and electricity: indeed, conceivably more powerful than any of those great technological revolutions of the past.

The lessons of history

Certainly, the experience of past innovations suggests that information and communications technology would have a transforming economic impact. Take the railways and electricity, two of the great innovations of the nineteenth century. Both were "disruptive" technologies – they radically altered the way people did things, from running businesses to managing their social lives.

The railways, the greatest transport revolution of the nineteenth century, had a striking impact on growth. The carriage of freight by rail added perhaps 10 percent overall to American output over a couple of decades in the late nineteenth century.[3] The best guess for the impact of business-to-business commerce is that it will have about half as much effect as the railways did, permanently boosting the level of output by an average of 5 percent in the rich countries, with more than half the increase appearing in the first ten years of the century.[4] The wider the benefits of the Internet spread, the greater the gains in growth are likely to be.

But the Internet works in concert with earlier investments. Computers, software, and telecommunications already account for about 12 percent of America's capital stock, or almost as much as the railways did at the peak of America's railway age, and investment in information and communications technology (ICT) has been increasing. Moreover, information technology may have more pervasive effects than did previous developments in physical transportation. For the Internet, in particular, is not just a new way to transport services. It creates a new marketplace and a new information system as well as a new means of distribution and communication.[5]

Among the lessons from the introduction of electricity is the need for patience. The dynamo, which opened the way for the commercial use of electric power, was introduced in the early 1880s.[6] Yet, by the end of the nineteenth century, electricity accounted for less than 5 percent of the power used in American manufacturing. Even in 1919, it accounted for only half. At that stage, thanks to Henry Ford, companies began to realize that electricity not only served as a source of power, but enabled machines to be used quite differently. Machines that had previously been placed next to water wheels or steam engines could now be placed along a production line, thus maximizing the efficiency of the work flow. Because it was possible to embed an electric motor in a machine tool, it also became possible to arrange the stages of production in a logical order and to remove the cumbersome belt-drives that had previously carried steam power around a factory.

. . . . .

The adoption of both railways and electricity was speeded by the effect they had on costs. Between 1870 and 1913, rail freight rates across America fell in real terms by an average of 3 percent year; between 1890 and 1920, electricity prices dropped by an average of 6 percent a year. However, the plunge in the cost of electronic communications has been faster than anything seen in previous technological revolutions. Over the final three decades of the twentieth century, the real price of computer-processing power dropped by an average of 35 percent a year; since 1930 the annual fall in the cost of a three-minute telephone call from New York to London has been 10 percent. No wonder the Internet is being adopted so much faster than those two previous general-purpose technologies.[7] In addition, many more people are now aware of the potential of technological change to increase efficiency, and are hunting systematically for ways to make it do so.

Timing may also be important. The railways made much of their impact during the historical period that most closely resembles our own: the final quarter of the nineteenth century and the opening of the twentieth century. Then, as now, the liberalization of international trade and investment stimulated a long boom (although the restrictions on the movement of labor were less severe than they are today). The pace of technological innovation was racing forward, with the development of electricity, the telegraph and telephone, and the car. By contrast, electricity began to transform the productivity of the American economy only after 1920, just as the world was closing down, hemmed in by trade barriers and the brewing of another global war. On some views the Internet, because it coincides with and reinforces the world's second great era of globalization, could have an even greater impact on economic growth than did electricity in the 1920s.[8]

The productivity paradox

It seems self-evident that the communications revolution will transform productivity, just as the railways and electricity once did. As early as 1996 Alan Greenspan, chairman of the Federal Reserve Board, pointed out that, "The rapid acceleration of computer and telecommunication technologies can reasonably be expected to appreciably raise our productivity and standards of living in the twenty-first century, and quite possibly in some of the remaining years of this century."[9]

If one looks at the transformation under way in many companies, it is easy to agree with Mr Greenspan. In all sorts of ways, information and communications technology seems to be helping to cut costs and to use resources more efficiently. The Internet lowers search costs, it helps to shorten the supply chain, and it reduces some barriers to entry. It simplifies and reduces the cost of searching for information on potential suppliers, employees, and markets. By making it easier for firms to communicate with their suppliers about the state of demand, it ensures more efficient management of supply chains. By simplifying the management of stocks, it eliminates excessive inventory.

All this is easy to see from talking to individual companies but hard to demonstrate from economic statistics. Businesses have invested massively in information technology and communications. By the end of the century it was growing at an annual average rate of 12 percent – faster than investment overall – in the United States, the European Union, and Japan. However, productivity growth slowed sharply in most countries in the 1970s and 1980s. Only in the United States, in the late 1990s, were there signs that productivity growth might be speeding up.

What might explain this mystery? There are three possibilities. First, plenty of company spending on information technology has, in some sense, been wasted. Lots of people have computers on their desks with vastly more power than they will ever use (indeed, it is easier to waste time with a computer connected to the Internet than it ever was with a typewriter), and whole departments have been set up to manage office networks. No wonder many managers see information technologies as a bottomless pit into which immense sums of money vanish without trace.

A second explanation might be that economies can easily take a couple of generations to learn how to get the most from a new technology. Innovation tends to move through what is sometimes called the "reverse product cycle": gains in the efficiency of processes (such as getting customers to submit orders on-line) are followed by improvements in the quality of existing products (on-line share dealing), and then by the creation of new products. It is at this final stage that significant economic growth usually occurs. New products tend to beget other new products, just as Edison's electric lamp led to the development of power generation and delivery, which led in turn to vacuum cleaners, air conditioning, hairdryers, and the Sony Walkman.[10]

In the case of electricity, the big gains in productivity started to occur once it powered more than half of all industrial machinery. In 1984, computers were used, in some form or other, by a quarter of American workers; today, that figure hovers a bit above half and is lower everywhere else. But the IBM PC was not introduced until 1982, and the spread of the networked computer began only after that. With hindsight, these are probably more important landmarks. Indeed, the greatest opportunities for productivity gains are likely to result from the impact of the Internet itself, which hardly appeared on workers' desktops before 1996 – 97. If electricity is a good model, the surge in productivity will run through the first quarter of this century, bringing a long period of economic growth more rapid than anything seen in the rich world since the 1960s.

A third point to bear in mind is the sheer difficulty of measuring productivity improvements. Productivity statistics tend to measure change in the production of goods. They do not take full account of new goods that appear nor of the quality of goods. The value of many consumer durables – think of the PC itself – now lies in the software programmed into them rather than in the hardware itself. Productivity growth in services is even harder to assess – and yet service industries have been spending the most on information technology.

Measuring changes in the productivity of services involves solving two problems: measuring output and measuring quality improvements. With physical production, measuring output is easy. In the case of car manufacturers, you simply count the wheels that drive out of each plant. With services, the task is harder. Statisticians typically assume that the output of a nurse or a school teacher rises in line with the number of hours worked. Measuring inquiries handled over the telephone or the effectiveness of television advertising is more difficult still. Neither physical production nor time spent is a guide. Measuring quality improvements is also hard, even with physical goods. But while quality in an automobile might be measured in terms of reliability, an investment in improving the quality of a Web page is entirely in the eye of the surfer.

A result of these two characteristics – the immeasurability of output and the subjectivity of quality – is that some increases in productivity may actually show up as declines in output. If the Internet allows nurses to monitor from a distance whether patients are taking medication, the nurses may visit fewer homes to check on them. Their measured output may appear to decline, even though the quality of service has improved. Many other examples would show the same effect.

The largest service providers of all are governments, a fact that presents yet a third measurement problem. Many productivity increases may eventually appear in delivery of state services, such as welfare, tax collection, and education, but as these services are usually given away rather than sold on the market, their value is even harder to establish than that of, say, an Internet bank's Web site. In addition, because governments are notoriously late adopters of new technologies, productivity improvements in government services may be genuinely slow to take place.

Good for growth

Happily, in 2000 studies began to find evidence of the benign impact on productivity so frequently and authoritatively predicted.

In the final five years of the twentieth century, productivity growth picked up in the United States – although even that, according to a paper by Robert Gordon, an economist at Northwestern University, was entirely explained by two factors: the normal way productivity tends to rise when the economy is growing fast, and faster productivity growth in one solitary industry, computer hardware.[11] But investment in technologies such as the Internet and the Web browser only really got going around 1995, and it takes time for statistical trends to become apparent. Sure enough, in 2000, a piece of research by two economists at the Federal Reserve Board in Washington, DC, at last detected signs that information technology was the key factor behind America's productivity growth. They argued that roughly two-thirds of the acceleration in productivity growth between the first and second halves of the 1990s was directly due to the production of or investment in computers, which has been higher in the United States than in Europe's 11 "euro" economies or Japan.[12] [Fig 8-1]

· · · · ·

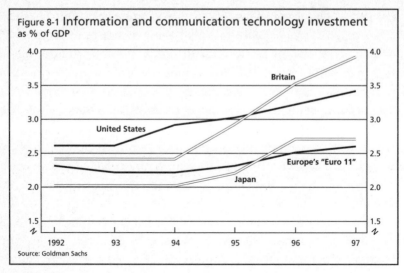

Figure 8-1 Information and communication technology investment as % of GDP

Source: Goldman Sachs

Evidence from companies (such as that described in Chapter 6) strongly suggests that this new research will turn out to be on the right track. Many began only in the final moments of the twentieth century to grasp the degree of organizational change that might be needed to take full advantage of the Internet. Work by two American economists, Erik Brynjolfsson and Lorin Hitt, found that companies received greater benefits from investments in information technology if these were accompanied by organizational change.[13] Companies do best if they have work teams that can run themselves independently; individual workers with lots of authority to take decisions, especially over their method and pace of work; high investments in training and education; and incentive systems that encourage team work. For a new start-up, it may be easy to build a company with such qualities. But to instil them into an existing company may mean replacing staff and abandoning business practices that have been successful for decades. It is a risky, expensive, and unpleasant undertaking for any company.

Evidence from other countries is still sketchy, but will surely tell the same story. Huge investments must be made, not in plant and machinery, which shows up in balance sheets, but in business processes and in recruiting the right employees and managers. All this takes time, and the rest of the world is starting well behind the United States.

Indeed, the main impact on productivity may turn out to be felt not in the United States but in other parts of the world, where corporate reorganization has made less headway and markets are less open and competitive. The cross-border uses of the Internet, which may turn out to be the most important, have hardly begun to develop. In September 1999 the number of Internet hosts for every thousand inhabitants in the United States was seven times that of the European Union and eight times that of Japan – and growing faster.[14]

Where distribution margins are highest, prices are likely to fall furthest and efficiency to increase most. In Japan, that nation of middlemen and interminable supply chains, the Internet has the power to give consumers a new voice. In Europe, with its rigidities and high taxes, the Internet will bring flexibility and transparency. In developing countries, where business has relied on contacts and special deals, the Internet offers an alternative source of information and supply. The real "new paradigm" is the potential for liberating consumers and unlocking growth in the world's most rigid markets.

The communications revolution almost certainly is raising productivity, especially in service industries, but in ways that are difficult to assess. A hidden boom is taking place, improving the quality of all sorts of services in ways that consumers are dimly aware of, but that the published figures ignore. These improvements enrich us, not by giving us more affordable material possessions, but in a more subtle and fundamental way. Improvements in services such as health care and entertainment will have more positive effects on our quality of life than improvements in material goods are henceforth likely to have.

Good for inflation

As well as higher productivity, the communications revolution seems likely to deliver lower inflation. Indeed, Lawrence Summers, Treasury Secretary under President Clinton, described the impact of the Internet as an oil shock in reverse.[15] In 1973 – 74 and again in 1979, sudden increases in the world price of oil by the oil-exporting countries delivered a traumatic shock to the world economy. The result was recession – the increases took between half and one percentage point off world growth over two two-year periods – and a rise in inflation.

The Internet alone will not kill inflation, which is essentially a monetary phenomenon. If central banks design their interest-rate policies to aim for the same inflation rate that prevailed before the Internet's

· · · · ·

arrival, then in the long run inflation will not alter. Instead, economies will produce more, and employ more people, at unchanged rates of inflation. But the early effects of the Internet will certainly exert downward pressure on prices.[16]

Part of the impact comes simply from the process of technical change, which drives down the cost of everything digital from PCs and telephones to software and telephone calls. But those cost reductions will affect prices only if there is enough competition. Luckily, the Internet reduces some barriers to entry, allowing new competitors into some markets. In addition, it becomes easier and less expensive for both companies and individuals to compare prices and shop around. As electronic markets spread, for everything from cars to labor to second-hand machine tools, economies start to approximate more closely to the perfect markets that economists write about but rarely encounter. When consumers are as well informed as producers, everybody who might conceivably be a buyer or seller takes part.

Such price comparison may turn out to be especially important when it starts to happen across borders. The introduction of the euro will encourage price harmonization in Europe and make it hard for companies to continue to charge widely differing prices in different markets. Elsewhere, global price comparison will also become easier as more of the world's currencies become pegged to the American dollar.

Some signs of that effect could be seen by early 2000. In Britain, a study carried out in December 1999 by Barclays Capital suggested that Internet prices were significantly lower than prices in high-street shops for a variety of goods (with discounts ranging up to 22 percent for books).[17] The greatest effect is likely to be on transactions between businesses, and especially on procurement costs. Here, work by Goldman Sachs concluded that possible savings from on-line purchasing, in industries that accounted for about one-third of the American economy, ranged from 2 percent for the coal industry to up to 40 percent for electronic components.[18]

However, the United States is already arguably the most competitive market in the world for many kinds of business. The implication is that, in time, the effect of the Internet on prices (and on other parts of the economy) around the world may be much greater than it will be in America.

Bad for profits – even dotcom profits

Belief in a new economic paradigm helped to feed the huge surge in Internet stock prices in the late 1990s. At the peak of the boom, in 1999, half the record number of 510 initial public share offerings in the United States were by companies in Internet-related businesses.[19] To many such businesses, investment capital seemed almost "free," stimulating risk-taking and experimentation to an unprecedented degree.

For a brief period, dotcom companies soared to astonishing heights. In 1998, the shares of Amazon.com rose 966 percent in value; the following year its market value climbed to more than $30 billion. In other words, a loss-making electronic bookstore was valued at more in terms of market capitalization than Texaco, a giant oil firm. The performance of some small start-ups was even more amazing. In early 1999, broadcast.com, for instance, had a paper value of almost $5 billion on revenues of $16 million.

What made such valuations all the more extraordinary was that hardly any Internet companies had made a profit. Even among the ten largest, only five were profitable in mid-2000 (among their number were several that supply services to users of the Internet, such as Microsoft and VeriSign, which owns Network Solutions, a company that registers domain names). Pointing to the challenges to Amazon's margins, one commentator remarked, "Amazon could yet become a $10 billion business with the profits of a corner shop."[20] Indeed, some Internet start-ups had marketing budgets larger than their entire revenues. Dotcom investors told themselves that they were taking part in a land grab. There might be room on the Internet for only one or two Web sites for book stores, or for only one or two search engines or electronic marketplaces. Thus it was worth a massive investment, they argued, to gain "first-mover advantage."

That logic, always questionable, looked more doubtful as big firms with established brands began to exploit the Internet in more coherent ways. Many early Internet companies, and especially those selling to consumers, consisted of little more than a brand and a good (but easily copied) idea. They worked on the assumption that a land grab would bring a permanent first-mover advantage. Hence their huge spending on marketing. But expensive advertising is a less sure way to lock in the loyalty of consumers than a proprietary technology (such as Microsoft's

. . . . .

Windows operating system) or an extensive network (such as the networks that mobile-phone companies have built). The very success of
the Internet is based on open standards, making it easy for new competitors to enter markets. As a result, a large share of on-line trade in a
particular commodity is likely to enjoy lower profit margins than a large
share of the same business would in the physical world.

Stock-market booms of this sort have accompanied earlier bouts of
innovation. The revenues of the American radio industry shot from $60
million in 1922 to nearly $850 million in 1929. Meanwhile, the value of
a share in Radio Corp of America, which sold radio sets, went from $5
to more than $500 in a stock-market boom that was fed by the unprece-
dented growth in the number of Americans with the means and confidence to play the market. After the 1929 crash, it took three decades for
the value of RCA's shares to return to their 1920s high.[21]

When the boom ended, the broadcasting industry remained, however. By 1928 RCA had launched the first nationwide broadcasting
service and was branching into two other nascent entertainment
industries, phonograph records and talking pictures. The boom in dotcom shares will be forgotten, but it will leave a technology refined, disseminated, and put to use at extraordinary speed.

As for profits in general, they may not rise as a share of the faster economic growth that the communications revolution brings. Indeed, better and less expensive access to information reduces the advantages
that buyers usually enjoy over sellers (think how hard it can be to compare prices of used cars, for instance). By putting an unparalleled
amount of information in consumers' hands, the Internet exerts enormous price pressure on the inefficient, the greedy, and intermediaries of
all sorts. That will eat into profits. Ignorance is the friend of fat profits,
and ample information its foe. This is a revolution that, like most past
technological revolutions, will bring benefits mainly to consumers, not
producers. "For society, the Internet is a wonderful thing, but for capitalists, it's probably a net negative," punned Warren Buffet, chairman
of the Berkshire Hathaway investment company.[22]

A New Economic Structure

• • • • •

Apart from its effect on productivity and profits, the other big impact of the communications revolution will be on economic structure. The shape of economies is already altering as technological change speeds the arrival of new industries and new areas of trade. Material goods account for a smaller and smaller proportion of world output: agriculture and industry together were only 38 percent of world output in 1998, compared with 45 percent in 1980.[23] At the same time, the steep fall in the cost of distributing information on-line is altering the economics of location. It is affecting where industries are concentrated, the balance between country and town, and the workings of the job market. It will reduce some of the disadvantages that far-flung places have suffered in the past, and bring new opportunities to developing countries.

The economics of location

The factors that determined where economic activity took place were once simple. Where land was rich, people farmed it; where coal was plentiful, they mined it; where water provided a source of power, they built mills and factories. Now, a rising share of the rich world's output has become "weightless," to use a phrase coined by Danny Quah, an economist at the London School of Economics.[24] So how, in an increasingly weightless world, do companies decide where to produce?

For companies that can distribute their output on-line, one might perhaps assume that location would become irrelevant. Given a decent communications infrastructure, an educated work force and a reliable legal framework, any product could be produced and sold from anywhere on earth. For some industries, the Internet has clearly brought a new freedom. "In a world of information technology, it doesn't matter where you are," say their bosses.[25] Some of the fastest growing new industries – call centers, data processing, direct marketing – have moved to places far from the main concentrations of economic activity. They have been attracted by relatively low wage costs, and flexible work forces.

Because companies have more freedom to locate, they can reduce costs and use resources more efficiently. In the four years after 1996, America's corporate relocation rates roughly doubled, to eleven thousand moves a year. Many of those moves were to parts of the United States where the labor market was less stretched and pay rose more slowly: the mid-west, Maine, Pennsylvania, and Kentucky. That trend almost certainly helped to hold down company pay bills, and so inflation, during the boom years. The transformation of communications undoubtedly made such relocation easier. Thus Rayonier, a big wood and pulp company based in Stamford, Connecticut, moved to Jacksonville, Florida in 1999, to be nearer its timber mills and to escape the high office costs of the north-east. The company found that communication was no harder from Jacksonville than from Stamford.[26]

In some parts of the world, communications offer a new competitive advantage: time zones. It becomes possible to take advantage not only of geography but of the full twenty-four hours of the world's working day. Again, that means more efficient use of global resources. Thus Bangalore, a city in south India, has a flourishing software industry. Its exports, two-thirds of them to the United States, were 63 per cent higher in the six months to September 2000 than in the equivalent six months of 1999, and accounted for an astonishing 12.5 percent of all Indian exports over that period. One of the fastest growing parts of the industry, which is expanding twice as fast as the American software sector, is "remote maintenance" – whereby Indian companies repair software for companies in other parts of the world, often taking advantage of the time difference to offer overnight service. Most clients of Infosys, India's leading software firm, are in the United States. At the end of its working day, its American office e-mails Bangalore with customer problems, and Bangalore technicians solve them while the Americans sleep.[27]

Increasingly, services can be provided or monitored from afar, using scarce skills more efficiently. Liebherr, a German company that is one of the world's biggest makers of tower cranes, provides an example. The company uses "teleservice" to program cranes globally from its main crane plant in Biberach, south Germany. Oskar Frech, one of the world's leading makers of die-casting machines, offers another. Its latest generations of equipment have systems that allow Internet access by the company's own engineers, even if they are half a planet away

from the machine. In both of these cases, Germany is learning to combine communications technology with the extraordinary engineering
skills of its *Mittelstand* companies – largely privately owned businesses
in a host of specialist niches – to build a global services business from
the heart of Europe.[28] Italians, also strong in engineering niches, are
doing the same. The London Eye, the spectacular Ferris wheel built on
London's South Bank to celebrate the millennium, has a cable-tension
system that is monitored by a firm in Italy, using a modem to fire data
back to a PC.[29]

Yet distance still affects location. The communications revolution has
created opportunities for companies to site themselves far from markets. But it has not overcome the powerful centripetal forces that create
clusters of similar businesses. In international finance, one of the most
weightless of all businesses, the distribution of activity has narrowed
rapidly as electronic communications have expanded. International
securities trading is now consolidating in three or four centers around
the world. Even dotcoms huddle together. Silicon Valley is one enormous cluster of such businesses; Bangalore another.

Indeed, the experience of Bangalore's software industry shows that
the death of distance has limits. Its exports are dominated by services,
such as custom software work, rather than the more lucrative business
of software products, in the form of packages. Indian companies could
develop packages more cheaply than software houses in Europe or the
United States, but their distance from those markets makes it hard to
keep up with changing needs and standards.

Economists, most of whom have long ignored or despised economic
geography, are now taking a fresh interest in it.[30] Clearly, the Internet
produces a radical decline in transport costs, and not only in the costs
of transporting a company's product. Employees also face costs for
travelling to and from work. Against such costs there need to be set the
benefits that still come from locating a business in the midst of a cluster of similar companies. A large population and an agglomeration of
businesses will guarantee a large pool of labor, and so give employers
more choice, especially if they want to recruit workers with specialized
skills. And there is, undoubtedly, a further benefit: the powerful way
that physical proximity seems to allow people to reach agreements,
think up new ideas, and cement bonds. In these ways, distance is far
from dead.

. . . . .

The effect on manufacturing and services

Few changes in industrial structure have been as striking as the convergence of manufacturing and services. The communications and computer revolutions help to blur the differences further. Three distinctions were important in the past. First, manufacturing produced a tangible object; services did not. Second, manufacturing operations could be at a distance from the final consumer; service operations could not. Third, manufactured goods could be mass-produced; services had to be individually created. Now, manufacturing is becoming increasingly intangible, more manufacturing is tailored to the individual's tastes, more services are being produced at a distance, and more services are being mass produced.

A growing share of the value of many physical products already comes from intangible aspects such as design and marketing. Tangible goods have more and more knowledge embedded in them. Intangible inputs, such as software, now account for 70 percent of the value of an automobile, for example.[31] A statistic in a 1996 speech by Alan Greenspan, chairman of the Federal Reserve Board, underlines the point. America's output, measured in tons, remains about as heavy today as it was a century ago, even though real GDP, measured in value, is twenty times greater. This much quoted passage may be hard to prove. Nonetheless, it draws attention to the growing importance in economic output of "knowledge" – design, styling, advertising, marketing, selling, consulting, and advising.

In addition, more and more manufacturing is customized. No longer will Henry Ford's principle of customer choice – "any colour you want as long as it's black" – dominate manufacturing. Instead, the technique developed most highly by Dell Computer is now used in many other industries, and in business-to-business applications as well as retail ones. Thus idtown, a watchmaker based in Hong Kong, offers on-line customers an almost infinite variety of designs of watch, assembled from standard parts, for much the same price as a mass-produced watch. The highly individualized production process and the interaction with the customer is more typical of old-style service industries than of mass production.

At the same time, many service industries have acquired characteristics once associated with manufacturing. Low-cost communications permit many services to be mass produced. As long as retail banks needed clerks to sit all day at counters dealing with individual customers, a bank's growth was limited by the size and number of its branches, which in turn constrained the number of customers who could be conveniently served. With banking transactions being handled by customers who push buttons on an automatic teller machine, a telephone, or a computer keyboard, banks can serve larger markets, both in terms of geography and of customers per employee.

With services as with manufacturing, companies can now combine economies of scale with personalized service. Inexpensive access to sophisticated databases enables businesses such as banking and life insurance to tailor both their marketing and their products to the specific needs of individuals. Services resemble manufacturing in another way: they can increasingly be produced at a distance from their final market. Financial services, entertainment, education, security monitoring, secretarial services, accountancy, and games can all be produced and sold at a distance from the ultimate consumer. Such long-distance provision is possible even in medicine, where some services no longer require doctor and patient to meet face to face. Once, a blood test meant a trip to a clinic and a disagreeable session as a nurse or a phlebotomist inserted a needle and filled a vial with blood. Now, two companies working jointly, one American and the other Japanese, have invented a gadget that, when placed on the skin, can read the broad chemical content of blood.[32] Since such information can then be stored in a computer and transmitted electronically to a doctor anywhere in the world, the two key aspects of a service – immediacy and the need for personal contact – may vanish, even from some kinds of medical care.

In future, one of the most important distinctions between products and processes may be, not whether they are goods or services, but whether they still require physical distribution or are truly intangible and so capable of being delivered on-line. Products that require physical distribution may continue to enjoy a gradual decline in distribution costs, partly because of changes in freight technologies but also because electronic communications will improve the efficiency of physical distribution. Products that are weightless, though, will enjoy a larger and faster fall in distribution costs.

. . . . .

The effect on trade

Strikingly, some of the services that are growing most rapidly in many countries are hardly traded at all. Of America's fastest growing service industries – telecommunications, health and medicine, education, insurance, travel, and communications (including movies, recorded music, and television) – only the last three are widely bought or sold abroad. American exports of health and medical services, the country's largest single industry, are tiny.

Rich countries have made big investments in education and medicine that may give them a comparative advantage over less developed countries. Already Britain's Open University runs, by mail, telephone, and the Internet, Europe's largest business school. Other universities are emulating it. In Alaska, a dedicated medical telephone network links local health workers in more than one hundred villages with doctors at a regional hospital. If such a network can be established within Alaska, why not between Alaska and, say, English-speaking countries in the Caribbean?

So some products, which have traditionally not been traded between countries, may now come to be exchanged. But others, which have been traded, may now have a new reason to be produced nationally or even locally. The process of mass customization may increase the amount of production that takes place on the spot or at least not far from the purchaser. Express Custom Tailors, an American company established to produce made-to-measure menswear, ordered electronically, has located its main plant not near the sweatshops of Vietnam or Guangdong, but in Cleveland, Ohio. The reason is partly that the town's long tailoring tradition gives it a ready supply of trained workers; and partly that, by producing in America, it can shave precious days off delivery. Those extra moments make sense, given the company's aim: "a custom garment in less time than you have to wait for the alterations, at a lower price than off-the-rack." They suggest that, unexpectedly, companies in industries under attack from low-cost foreign competition may be able to harness communications technology to move upmarket, taking advantage of speed of delivery and the superior ability to control inventory that the Internet allows.[33]

Distribution costs will thus remain one determinant of location for some kinds of industry. For others, the deciding factors will include reg-

ulation, infrastructure, economic and political stability, the education
and culture of the work force, language, and time zone. That last factor
may benefit Europe, the only one of the three main continents in a time
zone that covers part of the working days of both the other two. Companies will still want to cluster with others in the same business. But
many clusters will emerge as the global or national center of a particular trade, not merely the local center. Their market will be a continent,
a region, or indeed the whole world.

The effect on jobs

Despite such reassurances, many in the industrialized world worry that
production and jobs will shift on a terrifying scale from the rich world
to the poor. People in white-collar jobs in rich countries fear that their
work will vanish down a telephone line. They see themselves exposed
to the competitive pressures that have already squeezed manufacturing
workers.

Certainly, some activities will migrate from rich countries to poorer
ones, and the most footloose industries will move away from the richer
industrializing countries. Basic data processing, for instance, is already
shifting to China and Vietnam. But, where labor markets are reasonably flexible, new jobs will appear. The problem is, when old jobs go,
everybody notices. When new industries spring up, they often do so
unannounced. Pessimists easily forget that, during two centuries of
immense technological progress, employment has risen almost contin-
uously. Millions of jobs have been destroyed; millions more have
sprung up to take their place.

Many jobs will emerge through the export opportunities opened up
by low-cost communications. As John Heilemann, an American journalist, puts it, "More Americans make computers than cars; more make
semi-conductors than construction machinery; more work in data processing than oil refining. Since 1990, US firms have been spending
more on computers and communications gear than on all other capital
equipment combined. Software is the country's fastest growing industry. World trade in information-related goods and services is growing
five times faster than in natural resources."[34] Heilemann could have
added that, since 1990, America's movie industry has created more jobs

than its automobile manufacturers, pharmaceutical firms, and hotels combined; that the computer-software industry employs ten million people around the world; and that call centers, an industry that barely existed in 1990, now employ four million people in the United States and more than a quarter of a million in Europe.

All of these jobs require skills such as articulacy, courtesy, creativity, accuracy, and resourcefulness, which schools and universities need to help their students develop. These skills are not specific to gender, age, or race. So not only will opportunities be greater, but more people will also be able to compete for them on equal terms.

If people are flexible, and if government regulations do not inhibit entrepreneurs, new jobs will arise, often in industries that did not exist a decade or two previously. But, just as the pattern of employment alters, so will the pattern of pay.

Distributional effects

The death of distance will affect incomes in ways that at first seem paradoxical. Between countries, incomes will become more equal; within countries, they will become less equal. Everywhere, the premium for skill, creativity, and intelligence will rise.

Indeed, it seems that, while the coming of mass production increased the relative demand for unskilled workers, the coming of computers and communications raises the need for skills and education. Low-skilled clerical and production jobs are more readily replaced by computing power than are professional and managerial jobs. Besides, computers seem to enhance the value of what skilled workers do: they increase the return on using information creatively, whether to design a new product or to manage a company. Almost certainly, technological change accounts for much of the widening gap between people with college education and those who dropped out of education early or only completed high school.

The drive to recruit the best affects job markets in many countries. Big firms already scour the world for the rarest talent. At the top end of the pay scale, the Internet's powerful search capabilities are helping a global market for top-quality labor to emerge. When BA, Britain's largest airline, parted company with its chief executive in 2000, it found his replacement at Anslett, an Australian airline. At a lowlier level, most companies with a Web site include a "come and join us" page. Some specialist Web sites enable employees with particular skills – chief finance officer, for instance, or software designer – to compare what they get with the going rate for the job. The result is that the market for talent, like so many other markets, has become more open and transparent. In such markets, inexplicable price differences for similar products rarely survive.

Add to that the fact that wages, especially in industries producing goods traded internationally, tend to track productivity. One effect of the death of distance will be to increase the proportion of every economy accounted for by trade. Another will be to make companies more footloose – more willing to locate wherever the best combination of skills and productivity can be had. That too will tend to iron out differences between countries. Wages will be calibrated more precisely, in more industries, with a world standard of productivity.

These forces will tend to erode differences between countries, especially for the scarcest talent. Talent can move. At the same time, the discount now imposed by the economy for lack of skills will be driven down even further. And the premium for the top people in many occupations will remain high. In companies, the folk at the top benefited hugely in the late 1990s from the spread of stock options, a phenomenon driven partly by the boom in technology stocks. Conventional companies found it hard to retain talent as booming Internet start-ups lured away good people with stock options, so they responded with armfuls of stock options of their own. In addition, some gained from an effect described in a book by two American economists as the "winner-take-all society."[35]

Two simultaneous changes contribute to this effect. First, many more products, such as a piece of software or a video of *Toy Story*, cost little or nothing extra to reproduce millions of times. Second, inexpensive, fast communications widen enormously the market for many products. Buyers and sellers can now discover the global best, rather than settling for the local best. So a singer who would once have charmed a city now has the potential to conquer the world.

Thus, as electronic reproduction slashes the costs of making perfect copies, electronic distribution puts star performances in front of a larger market than ever before. A huge increase in demand for a product that was once in finite supply can now be satisfied at little extra cost. Result: the rise of the superstar. A few top performers, in sports, music, movies, and television, make stupendous amounts of money.

The superstar phenomenon will spread to other jobs. The market will pay lawyers, bankers, doctors not according to their absolute performance, but according to their performance relative to others in the same field. In those jobs offering the greatest rewards for being or having the best, the best will earn fortunes.

Growth and the Knowledge Revolution

· · · · ·

The death of distance transforms and also depends upon innovation. New technologies, new products, new ideas are spreading faster than ever before. It is less expensive to develop and launch new products, and easier to find potential customers and investors. The links between innovators and their financiers at one end of the chain and their customers at the other have grown shorter. The market economy, now adopted in almost every country on earth, will allocate resources more effectively on the basis of ever better information.

The late 1990s saw a surge in spending on research and development in the OECD countries, together with a shift in R&D away from spending by governments on defense and toward spending by businesses on market-related projects. To some extent, the innovation that has accompanied the growth of the Internet in the United States may be part of the "peace dividend." Spending has also grown in other knowledge-rich areas, such as education and software. Indeed, investment in these intangible assets is now as large as investment in fixed capital equipment such as machinery.[36]

Significantly, service industries, once largely uninterested in R&D, now account in several countries for a bigger share of it than does manufacturing. Service industries have traditionally been slower to innovate than manufacturing, but some are now more innovative than manufacturers and concentrate on making similar improvements, in quality, market, and range.

The pace of innovation has accelerated. Surveys in the United States suggest that a project's average R&D life has dropped from eighteen months in 1993 to ten months in 1998.[37] The pace may speed up further, because so many on-line products are readily imitated or copied. The Internet makes it easy for potential competitors around the world to watch one another. Monitoring innovation by others has become an important part of R&D in many firms.

Opening markets

Some of the effects of new communications will be felt most acutely outside the United States, and especially in the new industrialized countries. But they will spread faster if markets are open.

The global market in communications – not just in hardware, but in communications services – must continue to be liberalized. That has begun to happen under the auspices of the World Trade Organization, which in 1997 persuaded some seventy countries to open their markets to unfettered competition in telecommunications services. That is a big step forward. But it will need the sustained support of governments. Experience in countries where liberalization happened early – such as the United States and Britain – shows that competition needs active and energetic backing from government, not just dour acquiescence.

In addition, world trade in services in general must be set free. The argument for free trade in services is just as persuasive as the argument for free trade in goods: countries can make themselves richer by specializing in activities in which they have a comparative advantage, whether it be growing coffee or producing blockbuster movies.

But, while trade in goods was extensively liberalized in the second half of the twentieth century, all sorts of barriers hamper trade in services. It is harder to buy, say, health insurance from a foreign insurance company than to buy a foreign-made automobile or videocassette. And, while many governments pay overseas construction companies to build roads or power plants, they are less likely to pay for their citizens to consult foreign doctors or to learn from foreign teachers.

Liberalizing trade in services will not be easy. Barriers frequently stand in the way – not customs duties or tariffs, but domestic regulations and rules about qualifications and the like. Think of the obstacles to trade in domestic banking, medicine, legal services. Attempts to harmonize domestic rules may provoke fierce opposition: they provoked

American demonstrations against the World Trade Organization in Seattle in 1999 and Washington, DC, in 2000. Yet, without ways to harmonize such regulations, or dismantle them, international trade in some services will be held back.

In fact, the most powerful force for liberation may be communications itself. The more people can buy services on-line, the harder it will be to impose barriers to prevent them from doing so. Good communications will therefore drive liberalization, improve the quality of services all over the world, and extend consumer choice.

The Five P Problems: Policing the Electronic World

Ithiel de Sola Pool, one of the early thinkers on electronic communications and their implications for our lives, called them "technologies of freedom" and argued that they create an unprecedented opportunity for freedom of speech that should be reinforced rather than constrained.[1] So they do, but they also bring problems – problems that are not exactly new, but present themselves in new ways.

The main ones can, serendipitously, be grouped under the letter P.

- **Policing** is normally geographic in scope. It applies in cities, states, countries. So individual governments lack the power to regulate much of what happens on the Internet. Not surprisingly, many of the activities that first migrate on-line are those that want to escape regulation.

- **Pornography** is shorthand for the many disagreeable sorts of material that are easy to communicate on-line. Want to know how to make a bomb or launder money? Want to libel somebody or promote outrageous views on the Holocaust? The Internet gives you a platform. Sometimes, your freedom to say things to which others object will be defensible as free speech. But countries differ on what is or is not acceptable.

- **Protection** is essential to the vigorous growth of electronic commerce – of consumers, contracts, brands. Unless business has an environment as secure as the one it enjoys in the off-line world, it will not thrive.

- **Privacy** Electronic networks are the finest technologies ever invented for collecting personal data. And, once such data is stored, its sale is hard to regulate.

- **Property** The ideas, knowledge, and creativity that are the main non-human assets of most businesses in the electronic world are harder to protect from theft than old-fashioned drop-on-your-toe property. Besides, those who steal electronic property may not think of themselves as thieves. After all, they would argue, the owner still has the original.

Many of the rules that governments apply in the off-line world can be extended, at least in principle, to the on-line one. But often the rules are harder to apply on-line. One reason is that commerce, entertainment, and information now hop easily across borders. The private sector – and individual activity – are thus moving beyond the regulatory sway of the nation state. In addition, the democratizing force of electronic communications brings opportunities to many more people – including the opportunity to do something unpleasant or criminal. The individual has more freedom, just as Ithiel de Sola Pool predicted, but it is the freedom to do harm as well as good.

Policing the Electronic World

A borderless world is bad news for governments. Borders, after all, are what define their jurisdictions. The Internet makes it harder for governments to enforce their will, partly because it is international and so brings together incompatible jurisdictions, and partly because its technological characteristics make enforcement difficult.

A previous chapter looked at the way the Internet erodes governments' tax-raising powers, and argued that countries would have to work together to get things done. But there are many other areas in which economic activity on-line will ideally require rules that apply internationally, on issues such as privacy, security, and consumer protection. Yet countries often resent not being able to set their own standards. In Seattle in November 1999, American agitators demonstrated against a meeting of the World Trade Organization because they did not want the United States to be bound by international rules on labor and the environment.

But the limits to sovereignty set by the Internet cause other, more alarming problems. Not surprisingly, some of the first activities to migrate to the Internet have been those that governments try hardest to police, such as pornography and gambling. National laws are difficult to enforce, partly because tracing the source of material can be difficult, and so jurisdiction is often unclear, and partly because laws differ among countries.

Governing on-line material

As a vehicle for circumventing the law, the Internet is ideal. It can carry illegal material across international borders, covering its electronic tracks, and deliver it straight to the desktops of millions of individuals. Tracing its sources can become more difficult than it is when there is a paper trail to follow. Services can migrate to countries where laws are lenient or weakly enforced, creating offshore havens for pornography, gambling, and tax evasion, and breaching international rules on intellectual property.

In principle, the national laws that already exist to regulate such materials should simply be applied in cyberspace. When people break the law on-line in their own country, that is easy. If you libel your next-door neighbor on your Web site, you are in just as much trouble as you would be if you did the same thing in the local newspaper. But what if you live in another country, where libel rules are lax? Many problems arise where countries apply different standards.

For instance, in 1996, France asked the United States to crack down on an Islamic group in San Diego that had been putting on the Internet instructions for assembling inexpensive bombs similar to several that had recently exploded on the Paris metro. But the United States gives more generous protection to free speech than does France. American officials reportedly offered "sympathy" but little else.[2]

Intellectual property is often as hard to regulate as bomb making. One of the successes of the Internet has been as a marketplace for prescription drugs. By the start of this century, health appeared to have

. . . . .

overtaken pornography as one of the main reasons why people logged on to the Internet.[3] The advertising and selling of medicines is easier than in the real world, where patients have to trudge to the doctor for a prescription. But it raises at least three issues of policing. First, plenty of charlatans sell their wares on-line. Their goods may be ineffective, unapproved, or simply dangerous. When customers find that their on-line order for shark-fin cartilage fails to cure their cancer, they can report the offending site to a regulator, such as America's Food and Drug Administration. But if the site operates from another country, the FDA is unlikely to be able to track down, let alone discipline, its proprietor.

Second, even where a drug is legal and approved in one country, it may be banned in another. But pills are light and easy to mail, and when pharmacies sell by mail order to patients abroad, they may have to rely on their customers to order only what they are legally allowed to take. Several legitimate pharmacies in the United States will cheerfully mail prescription drugs such as Viagra, sales of which are tightly regulated in Britain, to British customers who reply correctly to a few cursory questions.[4]

Third, and trickiest of all, are differing rules on advertising. Pills at least need the cooperation of the postal service or a courier company to make their illegal journey across borders. Advertisements do not. Multinational drug companies are allowed to promote themselves and their products in the United States. But in many other countries, including the whole of Europe, it is illegal to advertise drugs directly to consumers. Yet governments, at least in rich democracies, cannot control which sites their citizens look at. In May 2000, Britain's Medicines Control Agency, the industry's regulatory body, banned two Web sites set up by Biogen and Schering, which both manufacture beta-interferon, an expensive and (in Britain) little-used treatment for multiple sclerosis. Their sites were deemed promotional, because they campaigned for greater access to the drug. Not surprisingly, the ban was ineffective.

Given so many differences, a resolution passed by the World Health Organization in May 1998 is likely to go unheeded. The WHO called on governments to tighten controls on the quality of information about drugs on the Internet and to clamp down on cross-border trading of drugs through the Internet where such trade is illegal. It will take many years of cooperation among different regulatory agencies to put that into practice. Meanwhile, the problem will get worse. In 1999, the Customs service of the United States seized four and a half times as many packages containing prescription drugs as they had done the previous year.[5]

Self-regulation or no regulation?

One consequence of the difficulty of policing on-line activities may well be a rise in self-regulation. That is not, of course, a cure in cases such as that of those San Diego terrorists, where two parties are determined to do something illegal. But, where the problem is to exclude hucksters and or shark-fin merchants, industries will increasingly set their own rules and find ways to marginalize those who break them.

One possible model is the scuba-diving community's Professional Association of Diving Instructors, which franchises tests for new divers around the world. Its authority comes mainly from the quality of the training courses and testing materials that it designs. Reputable dive sites invariably insist that divers should have PADI qualifications or the equivalent. In addition, divers share a few common interests, including an overwhelming interest in their personal safety.

In the same way, the bottom-up consensual governance of the Internet may provide a model for some regulatory bodies. But such bodies, like the organizations that regulate other, more commercially successful sports than diving, also show the limits of this approach. Once political and commercial pressures build up, self-regulation may no longer be enough. Governments may have to accept that there is more need to agree on how to regulate everything from advertising drugs to publishing recipes for bombs than ever before.

Pornography

· · · · ·

In the mid-1990s Jacques Chirac, the French president, referred to the Internet, with a certain Gallic hyperbole, as "a major threat to humanity." The president was worrying mainly about the threat posed by yet another channel for American culture into the French home, but people in many parts of the world feel qualms about the difficulty of controlling what comes down the wire or over the air.

Among the largest industries on the Internet are two that many countries tolerate reluctantly, if at all: pornography and gambling. Although still small compared with their off-line competitors, both are tailor-made for the electronic world. Indeed, the sex industry (or "adult entertainment," as its promoters primly call it) has helped to pay for and pioneer many techniques of electronic commerce. The cinema, the videocassette (75 percent of films in the late 1970s were "adult" movies), France's Minitel (an early on-line information network that carries what the French call "messages roses"), premium-rate telephone services, the CD-ROM: all were used, in their early years, partly or mainly to carry erotic content.

The debate over regulating pornography in particular reveals the difficulties of setting rules for a border-hopping medium. The regulatory problems are bad enough with the telephone, which supports a $2 billion industry of pornographic chat.[6] Calls are often routed, lucratively, through foreign countries with relaxed rules: countries with few telephone numbers are especially prized, because the short numbers fool people into thinking they are making a local call. But at least this is an industry that has long been run by large, usually state-owned, operators, used to heavy regulation. When, in the mid-1990s, an operator of sex-chat lines transmitted from Guyana to Britain broke the rules by advertising its services in a family newspaper, Britain's regulator threatened to route all calls to Guyana through an operator. Guyana promptly closed the line.[7]

On the Internet, control is many times harder, and the technology of delivery increasingly sophisticated. In addition, the Internet has almost certainly extended the market for pornography, including the sort of pornography that most people feel should be unavailable anywhere. In the United States, in September 1995, a two-year investigation by the Federal Bureau of Investigation ended with the arrests of dozens of people for using America Online to distribute pornographic pictures of children and to solicit sex with minors. In Britain, police believe that the Internet has made child pornography available to people who a few years ago would have had no idea where to find it.[8]

The problems of drawing a line

Whenever governments try to censor what people should be allowed to see, though, they run into immense difficulties. Are pictures of naked women an art form when painted by dead Italians but pornography when photographed by live ones? Even cohesive societies with uniform religious or cultural values find it difficult to decide where to draw lines: for a community of users scattered through about two hundred countries, the task is infinitely harder.

Because the Internet remains as yet largely an American colony, American values tend to shape it. The United States is a rarity: a country with free speech embedded in its constitution. Moreover, the whole ethos of the Internet, with its childhood spent in American universities, has been to protect free speech. Many Internet stalwarts would argue that the freedom to argue and discover is worth defending at all costs, even if its corollary is the freedom to titillate or abuse.

But as the Internet moves into the home, and as its users grow younger, the popular view changes. The American Congress has made two attempts to protect young users by legislating. In 1996 it rushed through the Communications Decency Act as part of the telecommunications bill of that year. This broad and vague provision made it a criminal offense for a person knowingly to transmit "obscene or indecent" material to anybody under eighteen; to display such material "in a manner available to" a minor; or to allow use of "any telecommunications facility under his control" for such prohibited activities. In a court case in Philadelphia later that year, the law was thrown out as unconstitutional, a judgment upheld the following year in the Supreme Court. The Internet, the Philadelphia judges poetically decreed, represented "a never-ending worldwide conversation."[9] As such, it deserved the same light hand of regulation that touches private discourse.

A second attempt to legislate to protect children from unsuitable content may go the same way. The Children's On-line Privacy Protection Act, which took effect in 2000, demands that Web sites obtain parental consent before collecting personal information on children under thirteen, and give parents a say in whether their child's information will be disclosed to others. A series of court cases is likely to test whether this act is more compatible with the constitution's concern for free speech than its predecessor.[10]

· · · · ·

The problem of accountability

At the heart of the problem is the blurring of public and private responsibilities that the Internet encourages. Many countries deal with issues of content by distinguishing between the public and the private. Governments in democratic countries have usually applied different standards to what their citizens say to each other in private, in letters, or over the telephone – and what they say in books or newspapers, on the radio, or on television. There have thus always been two classes of intercourse: one private and largely unregulated and the other public and subject to rules about matters such as decency, copyright and, in some countries, political balance.

Now, the line between public and private is blurring. The same network can deliver a newspaper, a broadcast, or a private letter; the same terminal can receive all of these. What starts out as private – an e-mail, say – can move seamlessly into the public arena (by being posted on a Web site, for instance). As a result, arguments about all sorts of on-line issues – copyright, sexual harassment, indecency – quickly turn into debates about free speech.

Even with stronger rules in place, it will always be hard for governments to curb the material available on-line, unless they are willing to intrude into the privacy of individuals to an extent that most democracies would not tolerate. Much of the sinister stuff on the Internet is put there by individuals who want no commercial reward.

So governments try to pin responsibility on companies. In the worlds of books and newspapers, liability for publishing legally unacceptable material normally rests with the publisher. Who is the publisher on the Internet?

In the Internet's early years of commerce, courts and governments sometimes tried to pin that responsibility on Internet Service Providers (ISPs). In December 1995, a prosecutor in Munich, Bavaria, told CompuServe, an ISP, that about two hundred on-line discussion groups on sex-related topics violated German law. Because CompuServe had no technical way of tailoring its content, which was held mainly on computers in Ohio, for German subscribers, it had to apply the exclusion everywhere in the world. But German politicians took a rather different view of the on-line carriage of illegal materials. "Providers that only transport contents can't be made liable for these foreign contents," said

Edzard Schmidt-Jortzig, the German Federal Justice Minister. "That is the only logical consequence, because we don't punish the postal services for transporting letters with instructions for concocting Molotov cocktails, Nazi propaganda, or child pornography. Those punishable are those who send it – whether over the Internet or as a letter."[11]

Censorship of material transmitted on-line will continue to present problems. It is difficult even for dictatorial countries that want to prevent their citizens reaching other sources of information. The most determined users of the Internet will always be able to do so, if only by dialing into an ISP in a foreign country and bearing the cost of an international, rather than a local, call.

Self-censorship or no censorship

Technically, a certain amount of undesirable material can be blocked by the use of screening technologies. These are used by governments of countries such as Iran or Syria, to control what their citizens watch;[12] by companies that want to prevent their employees from looking at dirty pictures when they should be working; and by households keen to control what their children see. Because the same software can be used by dictatorial governments and innocent homes, its development has made civil libertarians uneasy.

The software has improved since the time in the mid-1990s when America Online unintentionally shut down a forum for discussing breast cancer because it mentioned breasts, and software from another company closed off access to the official White House site because of a reference to the presidential "couple."[13] A filter system being developed in 2000 by the Internet Content Rating Association will be able to take context into account, so that a filter might block erotic images but allow pictures of nude bodies in a medical text.[14]

Screening invariably carries a cost, in terms of information forgone. That may be some deterrent to governments that want to censor unwilling citizens. Families, censoring what their children can see online, may be more willing to carry the cost. Several companies market filters to screen sites, the equivalent of the "V" chip that filters television material, allowing parents automatically to block programs they do not want their children to see.[15]

Civil libertarians worry that one impact of Internet screening technologies may be that private censorship replaces more accountable censorship by government authorities. If search engines refuse to list unrated sites, or if libraries or employers decide which filters should be applied, some sites will become unobtainable, with no redress. It is easier to mount a legal challenge to censorship by government than by a private organization.

This dilemma is not entirely new. In the nineteenth century, Charles Edward Mudie's Select Library dominated the London book-leasing business. Its success rested on the "young girl" standard, which Mudie applied to ensure that none of the reading material would make a Victorian lass blush, or raise moral or religious doubts in the middle-class home. The more adventurous Victorian novelists hated this commercially imposed censorship; patrons (and their wives) relied on it.[16]

If there is a large enough market for pre-screened material, then the demand will be met. But screening all available sites on the Internet is an immense task – much larger than that facing Mudie. Ultimately, the only sure way to control what people see will be by opting in, rather than leaving out: by restricting what can be viewed to a limited number of safe sites or approved television channels. That may be a large sacrifice for a family to make. But ultimately, the most effective censorship is likely to be that which families impose on themselves.

Protection

The Internet is a great place for fraudsters. Like any settlement on a new frontier it is heaving with outlaws and cowboys, protected, whether they are buyers or sellers, by the cloak of electronic anonymity. Even when traders are honest, on-line purchasing may leave consumers with less legal protection, and redress that is less readily available, than they would have if they were transacting in the off-line world.

Plenty of electronic fraud is there for consumers to buy into: a large number of sites offer fake identification, for example. Susan Collins, a Republican senator from Maine, successfully ordered on-line a US Army Reserve identification card, a Florida driving license, two press cards, and a Boston University student identification card, all for about $50. One fraud-tracking official in Florida claimed that, of the fake identities he saw, the proportion ordered on the Internet had risen from one percent in 1998 to 30 percent in 2000.[17]

Even without fraud, on-line consumers face problems, especially if they buy across borders. In a study by Consumers International, a consumer group, for the European Commission, researchers in eleven countries each ordered a common list of eight products on-line, aiming to buy where possible from sites in other countries, and then returned most of the items bought.[18] In total, testers ordered 151 products from seventeen countries. The project highlighted several problems. It was sometimes hard for a consumer to establish the identity of the company that provided a site, because the Web address did not correspond with the company's name. It was sometimes hard for the consumer to discover in which country a company was located. And both might change bewilderingly halfway through a transaction. For instance, an order for chocolates from café_tasse.be, a Belgian site, was re-routed to chocosphere.com, based in the American state of Oregon. In more than a quarter of cases, the retailer provided no geographical address. A buyer, however, might want reassurance on these points, if only to know where to complain if a transaction goes wrong and which jurisdiction the company trades in. Only 10 percent of the sites visited divulged which jurisdiction covered the transaction, and then it was invariably buried in the small print.

Not surprisingly, this test also revealed two other areas of concern: the difficulty of dealing with products that are ordered but fail to arrive, and of returning products once they have been bought. Fewer than half the test sites gave a target time for despatch and almost half of those failed to meet their target. Eleven orders simply failed to arrive. Many more arrived late or cost more than the consumer expected. Returns were also often problematic. Many companies added extra charges and some took an extremely long time to pay refunds, including a hundred days in the case of a pair of jeans ordered from Canada and more than 138 days with a Teletubby ordered from Australia. Any product purchased on the Internet is bought unseen; that, along with the problems that many on-line firms have with fulfilling orders, means that good policies on returns are important.

How should customers (and companies) be protected? One certainty is that those who want to buy and sell on-line across borders will have less legal protection than they enjoy if they shop at home. Governments can make various rules to improve the security of on-line contracts, such as agreeing to give legal force to digital signatures and telling Web sites to specify the jurisdiction that applies in case of any lawsuit. But there will be a trade-off between the amount of protection that consumers can be given and the extent to which transactions become cumbersome and inefficient.

Various intermediaries will help to transfer the business of protecting consumers from the public sector to the private. For instance, eBay offers discontented participants in its auctions an easy on-line arbitration service. Several cybercourts have sprung up, mainly in North America, to settle disputes on-line. They include eResolution, a Canadian company that specializes in resolving disputes about domain names; and ClickNsettle, a company based in New York that has long specialized in arbitration on insurance claims, but moved its operations on-line in the expectation that on-line arbitration would become increasingly popular. In Europe, where cross-border trading is likely to grow faster than it will in the United States, the European Union has been particularly anxious to promote the development of on-line dispute-resolution forums.[19]

In electronic retailing as in so many other on-line areas, the role of government will inevitably be curtailed. Either consumers will have to re-learn the law of caveat emptor – or private-sector intermediaries will spring up to provide some of the consumer protection that governments cannot offer. But they will do so only at a price. As a result, for on-line retail commerce to thrive, the benefits in terms of savings and choice will have to exceed the costs of do-it-yourself consumer protection.

Privacy

. . . . .

It was the spread of photography and inexpensive printing, and the consequent growth in media intrusion, that led Louis Brandeis, an American Supreme Court judge, to popularize, in an article in 1890, his famous phrase about "the right to be left alone." Communications technologies have been progressively undermining privacy ever since, and will continue to do so. The networked computer makes it easier than ever before to collect information that once was largely unrecorded, and then to store, retrieve, analyze – and sell it.

As a result people today have as little privacy as they did before the Industrial Revolution, when they lived in villages in which their neighbors knew their every move. Now, though, it is strangers who know all about them. The brief period of privacy that came between the end of village life and the coming of the networked computer is ended. Privacy previously survived by default: there was often no way to collect or analyze information. That passive protection has almost gone. Now, the protection of privacy will require deliberate decisions – by government, companies, and, above all, individuals. But handing over personal details often brings conveniences. Only when people regard the loss of privacy as a threat are they likely to be willing to accept the burden that privacy's protection, ironically, will entail.

The accumulation of data

The networked computer is the most efficient tool ever devised for collecting personal information. Everywhere people go in the electronic world, they leave behind a thick trail of data. Governments collect information on people as students, taxpayers, hospital patients, benefit recipients, law-breakers, and immigrants. Intelligence agencies from the United States, Britain, Canada, Australia, and New Zealand jointly monitor all international satellite-telecommunications traffic with a system called Echelon that can pick specific words or phrases from millions of messages.[20] America, Britain, Canada, and Australia are building national DNA databases of convicted criminals. It is probably only a matter of time before DNA is recorded at birth, to provide databases that cover entire populations.

While Big Brother is nosy, lots of commercial little brothers are nosier still. Companies squirrel away information from any transaction that uses a credit or bank debit card, from most financial transactions, telephone calls, and visits to Web sites. Increasingly, they will track people's whereabouts through their mobile telephones. In time, they will know what people do in their homes, because their domestic machinery will send back information on how it is being used and whether it needs servicing. Already, companies sell some of what they learn, and the trade in consumer information is vast and growing.

Americans are on more databases than any other people on earth, partly because they make more payments with a credit card than do people in other countries, and partly because the United States stores more information on more computers than less computerized nations do. One single company, Acxiom Corporation in Little Rock, Arkansas, has a database combining public and consumer information that covers more than 95 percent of American households and matches e-mail addresses to postal names and addresses.

Not only do companies scan their customers; they know more than ever about their employees. A survey carried out in 1997 by the American Management Association found that of nine hundred large companies, almost two-thirds admitted to some form of electronic surveillance of their own workers. Powerful new software makes it easy for bosses to record and monitor not only all telephone conversations, but every keystroke and e-mail message as well.[21]

The Internet adds enormously to the amount of information collected. Some is surrendered knowingly by users, as when sites offer free content in exchange for personal details. (Although such information is useful only if the provider is honest. One British user found that, by giving his birth date as that of the Duke of Wellington – 1 May 1769 – he was deluged with offers of special senior-citizen discounts from airlines.)[22] Other information is collected less obviously. Visit many sites, and your computer will be sent a "cookie," so that the next time you visit, you will be recognized (search for cookies.txt to see what is already on your PC). That file offers conveniences. You may be able to get straight into a site that would otherwise require you to remember a password, for example. But it also enables the site to track what you do when you visit. Each cookie contains a tracking number, allowing the computer to be identified on return visits. If you give your name during a transaction, or visit several sites that use the same tracking system, a cookie can cross-reference the information.

Most people were probably unaware of the amount of information they were disclosing until June 1999, when a row broke out over the purchase by DoubleClick, the world's biggest supplier of on-line advertising, of Abacus Direct, a database firm with 88 million names and addresses of customers gathered through direct-mail marketing. The deal brought home to many Americans the extent to which they could be tracked, analyzed, and sold. This came on top of an earlier storm of criticism that was unleashed when it emerged, earlier in the year, that Intel and Microsoft were trying to take one step further the PC's ability to signal its identity on-line. Intel planned to build identification numbers into its new Pentium III microprocessors in order, the company said, to improve security in electronic commerce; Microsoft was discovered to have a feature embedded in its Windows 98 operating software that sent hardware identification numbers to the company during a registration procedure. Both companies hastily offered software to allow users to turn off the identifying numbers.

But this is not the end of the story. More and more electronic devices and software packages contain identifying numbers to help them talk to one another. And new technology enables computers to identify, in a rough-and-ready way, the source of those who visit their sites by analyzing Internet addresses in one of the reverse directories (such as *www.arin.net*) that are readily available on-line. That will make it easier for some companies to target ads at specific users – and for others to keep some users away from their Web sites.[23]

Nor is it only when they use the Web that people sacrifice privacy. New technologies bring new intrusions. The mobile telephone, with its ability to track a caller's movements, has already proved a gift to the courts. Call records were used in the wrongful-death suit against O.J. Simpson. In Britain, records of calls to Malaysia and Indonesia, where match-rigging is rife, were an important part of the prosecution evidence in the 1997 trial of three footballers charged with fixing matches. Also used in that trial was data on where their mobile telephones (switched on, but not in use) were taken and for how long.[24] A nasty (although perhaps extreme) warning of the dangers that may come from the locating potential of mobile telephones was provided by the death, in April 1996, of Dzhokar Dudayev, leader of the Chechen separatists. He was apparently killed on a hillside while making a telephone call. He was hit by a missile fired from a jet after the Russians pinpointed his whereabouts by intercepting the signal of his satellite telephone.[25]

. . . . .

Increasingly, people will come to embody their identity, thanks to the advance of "biometrics," which will create inexpensive and reliable ways to identify people from their voices, eyeballs, thumbprints, or any other part of their anatomy. Already, for example, foreigners visiting the United States regularly can acquire an INS pass, which compares data stored on a plastic card with a reading of their right palms.[26] Combine this ability to identify any human being quickly and reliably with the ability of computers to store almost indefinitely a record of almost any action, and it is clear that the privacy people enjoyed during most of the twentieth century is gone.

The price of convenience

Much of this data collection brings not just commercial gains, but also benefits to individuals and society at large. Many of the ways in which companies use electronic communications, whether a telephone-call center or the Internet, to provide a more personal service rely on the ability to recognize a customer who pays a repeat visit. Good databases enable companies to screen out bad credit risks (and so, in theory at least, charge lower rates to good ones). When governments collect data, they generally use it for the greater public good: to tackle traffic jams, track fraudulent welfare claims, monitor book loans from libraries and, of course, catch criminals.

However, many people worry about the collection of so much personal information. A 1999 survey found that 87 percent of Americans were concerned about what happened to the information collected about them.[27] To many folk, the loss of privacy has unpleasant overtones of George Orwell's sinister fable of the future, *1984*. But are such worries justified? Peter Huber, an American communications expert, wrote a rebuttal of the idea that electronic communications permits government to threaten a citizen's privacy, called *Orwell's Revenge*. The sheer labor of keeping tabs on millions of citizens would deter Big Brother from doing so, he argued; and besides, maintaining a communications network requires the cooperation of too many skilled people for any totalitarian state to monopolize the task.

That may be a persuasive view when applied to established democracies with long traditions of respect for civil rights and the rule of law.

Indeed, it is no accident that the countries where copious electronic databases have been compiled have been those where governments are most scrupulous about protecting civil liberties. Americans part with large amounts of information about themselves because they are reasonably confident that it will not be misused.

But such views ignore two points: a government may want to keep track, not of millions of citizens, but of a few; and computers will readily spot the few in the crowd. In countries where civil rights are weak, interconnected electronic databases will make some aspects of government control easier. Even in Western democracies, there are real dilemmas when personal information collected for one purpose is used for others – when, for example, court records are available to lenders, or medical information to insurance companies. By cross-referencing information stored on databases, governments or companies may learn too much about an individual citizen.

Worse, information may fall into the wrong hands. Databases may be hacked into, or search services allow ordinary people to discover large amounts of information about others. A furious row blew up in 1996 over P-Trak, a service launched by Lexis-Nexis, one of the biggest providers of on-line data, and marketed to lawyers as a way to trace litigants. P-Trak initially allowed access to an individual's name, social-security number, telephone number, current and two previous addresses, and month and year of birth.[28]

In this case, the service produced a wave of complaints. But people continue to hand over personal information by the bucketful. Are there ways to protect privacy – and are people willing to put up with the restrictions that most of them entail?

Protecting privacy

Faced with worried consumers, the initial American response was to encourage self-regulation. Incidents such as the row over DoubleClick caused attitudes to shift. More than 80 percent of Americans say they would like government regulation of corporate use of personal data, according to a study by Odyssey, a market-research firm, in April 2000.[29]

The Federal Trade Commission, which had initially supported self-regulation, eventually grew exasperated with the failure of most Web sites voluntarily to develop adequate standards for privacy protection. In 2000, it reported that only 20 percent of sites applied what it regarded as best practice: giving people notice that data is being collected, choice about how such data is used, access to it once collected, and security to protect it from others. Several sites, including Infobeat, an Internet newsletter, and ReverseAuction.com, an auction site, were in trouble with the FTC for sharing data they had promised not to share. Studies of members of America's Direct Marketing Association by independent researchers found that more than half did not abide even by the association's modest guidelines.[30] In 2000 the FTC changed its approach and began to argue that legislation was needed. Many American states have already begun to pass on-line privacy laws, creating an inconvenient patchwork of different rules.

The EU has always had a regulatory, rather than self-regulatory, approach to privacy. It passed a directive on the protection of personal information which came into force in October 1998. Its aim is to give people control over their data by requiring "unambiguous" consent before a company or agency can process it, and by forbidding the use of data for any purpose other than that for which it was originally collected. The directive also bans the export of data to countries that do not offer equally strict protection.

As of 2000, few European countries had passed the domestic laws needed to give force to the directive. The measure's main effect had been to provoke a trade row, because the EU threatened to stop data exports to the United States. Faced with a prospective nightmare – American companies would have been unable to call up from their headquarters data on staff employed in Europe – a compromise emerged in March 2000. American companies, when dealing with European citizens, must observe the stringent EU rules in order to obtain "safe harbor" (as the diplomatic jargon has it) from prosecution.

In Europe, where so many Internet transactions cross borders, the desire for privacy protection is understandable. As personal details become an internationally traded commodity, people will begin to worry more about the issue.

Devising effective legislation to protect privacy will be just as difficult as devising all other legislation relating to on-line standards. In the interim, technology, united with the market, may offer some protection. Several Internet businesses offer to "remail" e-mail after stripping it of any identifying information. At least one Web site, *www.anonymizer.com*, offers anonymous Internet browsing. Software codes to encrypt commercial transactions offer customers some security – but no more than they would have if they mailed a physical coupon when they made a purchase.

Commercial services to help people protect their privacy will flourish only if people really want them. In the case of on-line transactions, consumers should at least be prominently offered a box to tick if they do not want their personal details to be passed on to other companies. They may frequently choose not to do so. In Britain, consumers are already offered such a box when they fill out forms with their personal details. Only a minority of form-fillers ever bothers to protect their privacy by ticking.

For the moment, most people in the rich world seem surprisingly willing to hand over their details, and often do so knowingly. That may alter as on-line transactions across borders increase, and as the cross-border trade in personal data grows. People may then begin to see their personal details in a new light: as their intellectual property, perhaps, to be protected as carefully as a valuable brand or a unique idea.

Intellectual Property

· · · · ·

The businesses that exploit the death of distance are built principally on intellectual property – intangible assets with commercial value such as brands, novel business techniques, good software, fine acting, captivating tunes, or good ideas. So, for companies, it is more important than ever to have robust ways to protect intellectual assets. A brand that is easy to recognize globally becomes enormously valuable. But, at the same time, an effect of electronic communications is to make intellectual assets easier to "steal" than ever before. Because it is cheap to copy and distribute them legitimately, it is also harder to stop people copying and distributing them without permission.

Both these points matter more to the United States than to any other country. Not only do American companies have by far the largest number of globally recognized brands (think of Coca-Cola, McDonald's, Disney), but America is also far and away the world's biggest exporter of intellectual property, thanks to its enormous earnings from the movie, music, and software industries. So America has an interest in global, rather than just national, ways to protect brands and prevent piracy.

Brands, trademarks, and domain names

As companies publicize their products globally on the Internet, they run into a problem that sounds abstruse but is actually fundamental to the use of global brands: domain names. The allocation of domain names is the power at the heart of the Internet. Easily the most valuable names, the Upper East Side of on-line real estate, are those ending in ".com." They are far more popular than ".org" or ".net", let alone tags denoting a country (as "de" and "fr" denote Germany and France). The popularity of these names is a striking symbol of American domination of the Internet. Although the country domain tag "us" exists, it is rarely used. It is an echo of the principle that Britain, where the postage stamp was invented, is the only country that does not place its name on its stamps. [Fig 9-1]

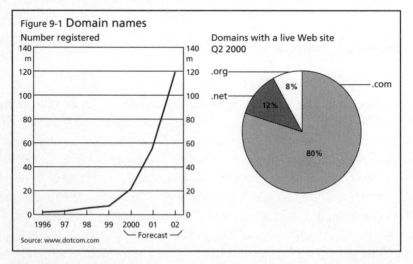

Figure 9-1 **Domain names**

Domain names are the Internet's equivalent of telephone numbers, and like telephone numbers they are unique. No two in the world can be the same. Given that, by June 2000, more than ten million domain names had been registered,[31] a memorable, readily identifiable domain name, designed to stand out in the sprawl of cyberspace, is clearly essential. Anybody looking for IBM, for instance, would rightly assume that its address was *www.ibm.com*. A domain name thus assumes as much importance to a company as a trademark. However, although trademark law allows many companies legitimately to share a name in the high street or telephone directory, on-line one name can serve only one company, because it is a unique locator.

That causes all manner of problems, because the desirable permutations of useful names are finite and rapidly running out. A survey in April 1999 found that, of 25,500 standard English-language dictionary words, only 1,760 were still free in the desirable .com domain.[32] So companies are forced to use acronyms or long and cumbersome words instead.

Even when a word is available, only one company can use it. Sun Microsystems happened to tie up *www.sun.com* before Sun Oil or Sun Photo. As John Gilmore of the Electronic Frontier Freedom Foundation once put it, "Neither lawyers nor government can make ten pounds of names fit into a one-pound bag."[33] As the Internet's commercial role grew in the mid-1990s, many giant companies found that their obvious domain name had already been snapped up. In the case of mcdonalds.com, an American journalist grabbed it before the eponymous hamburger empire – and then sold it back in exchange for a donation to charity. In other cases, companies have fought over a name. When Roadrunner Computer Systems, a small Internet company in New Mexico, registered roadrunner.com, it was challenged by Warner Brothers, owner of a cartoon character called the Road Runner.

An alternative is to use money rather than the courts to acquire a coveted name. Not surprisingly, the market for domain names is enormously valuable. One of the new registrars, register.com, reckons that it will be worth more than $2 billion a year by 2002. One name, business.com, has changed hands for $7.5 million.[34]

The resolution of disputes over domain names has become a pressing problem. Four accredited bodies deal with disputes, and one, the World Intellectual Property Organization, made an important ruling in 2000 on Internet addresses bearing the names of living people. The names of Julia Roberts, an American actress, and Jeanette Winterson, a British author, had been registered by cyber-squatters (people who register

. . . . .

Web addresses merely to sell them later), who had argued that the names of living people are not trademarks. The WIPO decided that the names should be returned to their rightful owners.

The Internet Corporation for Assigned Names and Numbers hopes to solve the problem of name shortage by creating seven "generic" top-level domain names, including .biz for businesses and .pro for lawyers, accountants and physicians. The problem here is partly that everybody sees suffixes other than .com – such as .net or .org – as second best. At the end of 1999, there were about 7.2 million .coms, compared with roughly 1.8 million .net and .org addresses.[35] In addition, the owners of trademarks have already paid heavily to acquire and defend their on-line names. They might have fought off or bought off cyber-squatters. Many would probably feel obliged to buy the new domain names as well.

Copyright and copying

With domain names, the intellectual-property problem is ultimately one with a legal solution: a set of rules must determine who has the right to each name. With copyright, the problems are more compli-cated. Not only are there legal issues of ownership, there are also tech-nical questions about how far legal rights can be enforced.

Much of the material bought and sold on electronic networks depends for its value mainly upon copyright. Yet copyright laws around the world were designed for an age when most copyrighted material was held in physical form – a book, say, or a CD. Once copyrighted material is converted into digital form, an infinite number of perfect copies can be made without any damage to or diminution of the qual-ity. The Internet amplifies that effect. It is, in a sense, one gigantic copy-ing machine, distributing copyrighted material by reproducing it.

That creates wonderful opportunities. The initial cost of developing software or recording a song may be high, but once it has been incurred, the product can be copied and distributed electronically many times at virtually no cost. But it also presents problems. For it means that when such products are sold, the price is vastly greater than the costs of producing each additional item. That in turn creates a huge temptation to cheat. Because it also costs almost nothing to produce a perfect fake, consumers and pirates share an interest in cheating the original creator of the product out of the payment that copyright law would normally guarantee.

Even before the Internet became popular, piracy plagued many intellectual-property industries. The music and software industries claim to lose billions of dollars a year through piracy, which in many parts of the developing world accounts for the bulk of sales. [Figs 9-2 and 9-3] The figures for lost revenue exaggerate: if inexpensive pirated software were not available, many developing countries would use much less. But they indicate problems so serious and so hard to control that the music industry in particular has long hesitated to sell directly over the Internet.

Figure 9-2 **Total estimated capacity for pressing CDs**

Country	Total estimated capacity, million units	Total legitimate demand for CDs, million units, 1999
Taiwan	3,900	190
Hong Kong	2,800	140
China	680	620
Singapore	490	50
Macau	340	negl.
Malaysia	280	50
Czech Republic	90	25
Russia	90	30
Israel*	90	9
Ukraine	70	5

Source: International Federation of the Phonographic Industry   *Not including Palestine Authority

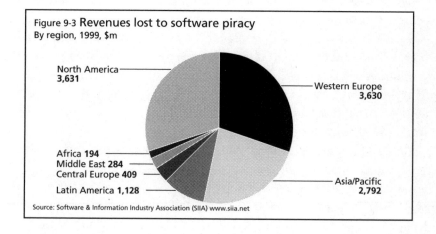

Figure 9-3 **Revenues lost to software piracy**
By region, 1999, $m

North America **3,631**

Western Europe **3,630**

Africa **194**
Middle East **284**
Central Europe **409**

Latin America **1,128**

Asia/Pacific **2,792**

Source: Software & Information Industry Association (SIIA) www.siia.net

Moreover, the problem is getting worse. In April 2000, the American record industry won a copyright case against MP3.com, a service that allowed people to download music from the Internet using a method of digital compression known as MP3. A few months later, the industry won another case against a service called Napster, which it charged with having "launched a service that enables and facilitates piracy of music on an unprecedented scale."[36]

Napster is the product of the development of a new breed of "distributed" systems that enable material to be stored on and retrieved from any one of millions of computers attached to the Internet. Because PCs have more disk space and computer power than was once the case, they can function like "servers," the powerful computers on which the material displayed on Web sites is generally stored. And because more and more PCs are permanently on-line, hooked up to the Internet via broadband connections such as cable modems or digital subscriber lines, they can be linked together to form a giant storage network.

In 1999 – 2000 this capability facilitated several services that allowed people to download music, not from record companies, but directly from files stored on one another's PCs. Napster, created by a nineteen-year-old American named Shawn Fanning, is not completely distributed: separate servers maintain central directories of available songs.[37] That made Napster vulnerable to prosecution for copyright violation. In autumn 2000 Bertelsmann, a big German music group, decided to take a big stake in Napster, hoping to turn it into a club of paying members, receiving music on-line. Another service, Gnutella, was created with no central directory of any kind. America Online, within whose network it was developed, ordered the program off its site a few hours after it had been posted. But by then, thousands of people had already downloaded it. Gnutella initially seemed too slow and not robust enough to be much of a threat. But others will refine it and develop new versions. The problems of the music industry will become those of other "digitizable" works, such as books, videos, and software.

Such file-swapping services will, if they grow, have many implications for the policing of the electronic world. No longer will illegal material be held on a central server that the courts can shut down, if they wish. Instead, the possibilities for lawbreaking will move further from the public realm to the private, and from the real world to the virtual. Illegal documents will have no fixed location. In that case, establishing a jurisdiction becomes all but impossible.

Paying for copyright

At present, people generally pay for copyrighted material each time they use it. They may still do so in future. But they may not pay as much each time as they did when such material was available only in physical form. As Ithiel de Sola Pool put it, "The question boils down to what users at a computer terminal will pay for it."[38]

Among the ways in which copyright holders might make money in a digital world is one proposed by Esther Dyson, an American cyber-guru. She has suggested that content could be sold outright to advertisers or sponsors who would give it away happily in order to attract attention to their own products. In fact, advertising has been the main way free music sites have made money – indeed, it has long been the way that most conventional forms of content, from newspapers to television broadcasting, are financed. Another model is to sell the razor cheaply but charge mainly for the blades. A book such as this one earns money chiefly from subsequent invitations to the author to speak at conferences.[39] A third proposal is what de Sola Pool dubbed "serviceright" – whereby a charge is made not for reproduction but for continuing service, in the form of, say, updating the original material. Not many people will bother to pirate a financial newsletter updated by the hour.

Even if a product is available free on-line, people may still choose to pay for it. Indeed, some evidence suggest that, even in the case of music, they already do so. People who sample music on-line appear to be likely to follow up by buying a CD.[40] In the first quarter of 2000, when Napster's popularity was at its height, American music sales were running 8 percent higher than a year earlier.[41] But people might also be willing to pay to receive music on-line.

Why might they do that? Largely because of convenience or quality of service. Napster and Gnutella built their notoriety on offering music free, making money only from advertising, but people might still pay to download faster and to have easier ways of locating precisely the piece they want. They might welcome a legal, commercial, universal jukebox, offered on a subscription or pay-per-song basis, with facilities to hunt for a song by genre or to receive personalized recommendations that the existing pirate services lack. In addition, most PCs may store the latest pop, but they are rather less likely to have that obscure 1960s track you want to hear (although the most popular piece downloaded from MP3.com in its early days was Beethoven's Moonlight Sonata).[42] And it is from their back catalogues that record companies make their largest profits.

The tale of the music industry does not suggest that it is impossible to make money from selling copyrighted material on-line. It does, however, point to the fact that the Internet encourages the growth of new models of distribution that provide users with what they want. If copyright owners are slow to create these themselves, plenty of clever nineteen-year-olds will do the job for them.

Indeed, if the new models are created, they may ultimately make more money for copyright owners, not less. MP3.com was the first company to notice the birth of a new market: its traffic peaked during the lunch hour, when office workers downloaded music to their desktop PCs. It may yet be that charging people a modest fixed sum to hear the music they want, when they want it, is more lucrative than charging them heftily each time they buy a CD that has some tracks that few people would pay to hear.

However, if the industry embraces a new model, the principal beneficiaries are likely to be the artists and the audience. The Internet offers a relatively inexpensive way to bring these two together, cutting out the record companies, or at least reducing their role. In all content industries, whether music, movies, or sport, the trend in recent decades has been for artists and stars to take a larger share of the revenue. The Internet will encourage that trend. Indeed, this threat to their importance may be a greater reason for the record companies to worry about new communications than the danger of digitized piracy.

Policing Global Networks

The death of distance, in a world of national jurisdiction, poses intractable difficulties for regulators. National laws simply cannot provide the means for dealing with many of the regulatory problems that the Internet, in particular, creates. At the very least, governments will have to cooperate more than they now do on subjects as diverse as taxation, terrorism, and Internet governance.

But even with cooperation, many issues will resist effective electronic policing, except at a cost that rich democracies will be unwilling to carry. The death of distance acquires a second meaning here: it becomes impossible to put distance between criminals and victims or between terrorists and governments as can be done in the physical world.

A necessary trade-off will result between policing and easy access to information. Just as in the non-electronic world, where societies choose to accept some risk in exchange for a certain level of personal convenience and freedom, so it will be in the electronic. Sometimes, the costs of effective control will be simply too high. In such cases, countries will have to adjust their expectations or levels of tolerance. So, for example, copyright holders will find different ways to earn revenue, learning to exploit the potential of electronic communications, rather than regretting the difficulties they cause.

Regimes that censor their citizens will also face a trade-off. They may be willing to accept more inconvenience and economic loss than rich democracies, but the costs will rise over time. In a world where, for the great majority, knowledge flows freely, the penalties for restricting it will be impoverishment and marginalization.

Knowledge and the New Monopolists

It is the most successful company ever created. It is probably the world's most valuable brand, having almost certainly overtaken Coca-Cola in the course of 2000[1]. For a while, its stock-market valuation was the world's largest. And yet Microsoft, whose commercial ferocity and acumen laid the foundation for American dominance of the new economy, was found guilty of a host of antitrust offenses and ordered by a court to split itself into two. For watchers outside the United States, that was a remarkable sight: what other country would treat a national champion so sternly? More significantly, Microsoft's case underlined an awkward aspect of the communications revolution. While it lowers many barriers to entry and creates new opportunities for competition, it paradoxically also seems to give rise to new monopolies.

Competition, clearly, does not come naturally in communications. In the new economy, many big companies dominate their respective markets. In hardware, think of Cisco, which bestrides the market for Internet routers, or Intel (microprocessors), or Oracle (databases). Market dominance is even more striking in the knowledge-based industries, whether software or content, that new communications foster. As of early 2000, three-quarters of all business-to-consumer e-commerce went through a mere five sites: Amazon, eBay, America Online, Yahoo!, and Buy.com.[2] Even when old-economy activities move on-line (book-selling, say, or auctions), a single firm seems to grab a larger share of the market than it could take in the off-line economy. In 2000 Amazon had about 80 percent of the on-line market for books in the United States.

The irony of this state of affairs is that the old world of communications was also dominated by monopolies. Indeed, in few other industries in the capitalist world was such a level of monopoly allowed and the spirit of competition so crushed. Significantly, one of the most famous antitrust cases in the United States before Microsoft involved the breaking up of AT&T, America's giant telephone company. Outside the United States, most countries had a single telephone monopoly, usually state-owned, up until the late 1990s, and often a dominant state-owned broadcaster too. Even when governments began to realize that competition in telecommunications was overwhelmingly in the national interest, it often proved hard to stimulate. In Britain, where competition began in 1984, British Telecom, the former state monopoly, still had more than 76 percent of the wired residential market and 49 percent of the business market in 2000.[3] In the United States, the Telecommunications Act of 1996, intended as a liberalizing measure, was followed by a surge of mergers which may, perversely, have increased concentration in some markets. The struggle to create effective competition in these older industries suggests the size of the task ahead.

The dilemma for regulators

Why is concentration such a common feature of these industries? Part of the answer is that they share certain characteristics that encourage it. In particular, they depend on networks, which rise steeply in value the larger they become. At the same time, many communications businesses have another peculiarity: their overheads and start-up costs are high, relative to the very low costs of producing each item they sell. The development costs may include hiring clever people or establishing a brand, but once they have been incurred, the product itself may be sold over and over again with hardly any extra cost to the company. Think of software or music or a page of information. Again, this is not new – it is typical of pharmaceuticals and movies – but it is especially common in the new economy.

The combination of networks with low additional costs of production means that a new-economy monopolist has different incentives from the typical old-economy monopolist. With physical goods, monopolists aim for scarcity. OPEC's oil exporters try to ramp up the price for their commodity by cutting back production. With software, the incentive is plenty. Microsoft has an incentive to get its product on to as many computers as possible, increasing the size and thus the value of its network. The more people use the software, the greater the incentive for developers to write new applications to work with it, and so for even more people to acquire it.

That brings enormous potential benefits to consumers. In the old economy, companies that expanded production quickly faced rising costs and declining returns. By contrast, when demand rises in new-economy businesses, so do returns, bringing greater efficiency and lower prices, stimulating yet higher demand. As the market for Yahoo! or Windows expands, their respective companies do not have to invest to build new plants as a car factory would have to do. They may thus enjoy rising returns to scale, the benefits of which will be shared with consumers.

But network effects have a more sinister consequence, say some. Once a product is established in the market, the demand for similar products will collapse. Consumers will be "locked in." Network effects thus foster monopoly.

Or do they? A converse view of the combination of high start-up costs and low marginal costs of production is that such businesses have a powerful need to avoid competition. Otherwise, competitors will speedily drive down the price at which the company sells its product toward the cost of producing it – zero, or nearly zero. As a result, argued Lawrence Summers, Treasury Secretary under President Clinton, the pursuit of monopoly power is "the central driving thrust" of the new economy.[4] In a knowledge-based economy, "The only incentive to produce anything is the possession of temporary monopoly power," which becomes "the central driving thrust of the new economy."

Sometimes, governments respond to such arguments by stepping in to protect new-economy firms from competition. When the assets of so many companies take the fragile form of intellectual capital, easily copied or stolen, they need government protection in the form of patents or copyright to survive. In 1999 the United States Patent and Trademark Office awarded 161,000 patents, almost twice as many as it had given out ten years earlier.[5] [Fig 10-1]

Figure 10-1 **Number of patents granted in the United States**

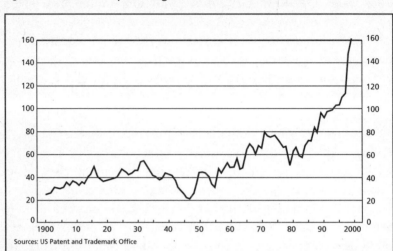

Sources: US Patent and Trademark Office

The key issue is whether protection is really the right way to stimulate technological innovation. Joseph Schumpeter, a Harvard University professor who was one of the pioneers of the theory of economic
development, argued that the protection that monopoly confers on
innovators might lead to a faster rate of progress than competition
would do. He coined the phrase "creative destruction" to describe the
process by which new companies innovate in order to take over markets
previously dominated by others. A classic example is the way that
IBM's mainframe monopoly, attacked in a long drawn-out antitrust
case in the 1970s that was finally called off by President Ronald Rea-
gan, was already crumbling as Apple Computer and others developed
user-friendly personal computers.

Whatever the temporary benefits of monopoly, competition is clearly
ultimately beneficial in both the old sort of communications industry
and the new. Market dominance is not in itself always a bad thing:
where companies dominate their market, they may not necessarily
abuse their position. But the temptation to do so will be greater. In the
lengthy argument over Microsoft, everybody readily agreed that Windows dominated the market for operating systems. The arguments
were about how Microsoft had used its clout: whether Big had also
been Bad, excluding new entrants, squashing competitors, and fleecing
customers. In the words of Joel Klein, the assistant attorney-general for
antitrust at the American Department of Justice who brought the

Microsoft case, "While technology changes, human nature does not...When it comes to antitrust enforcement, the new, new thing isn't so new after all."[6] The question for governments is how far they should allow competition, and how far they should protect the evanescent innovations of the new monopolists.

Are Communications Monopoly-Prone?
· · · · ·

It may seem odd that an industry in the throes of such a revolution as the communications business should be so full of market giants. It is even odder in the case of the Internet, which counts openness as one of its most striking characteristics. Indeed, the sheer number of dotcom businesses hardly suggests a technology that is naturally exclusive. Yet much of the language of the new economy – land grab, first-mover advantage, winner-take-all economy – is about the scramble to become the biggest or the only player in the field. Essential though competition is usually assumed to be for economic growth, the essence of the communications business is frequently monopoly.

The economics of networks

Communications and information technology are dominated by networks. These have a peculiar characteristic: the more points of access, the more valuable they are. Some even argue that this quality of communications networks can be formulated as Metcalfe's Law (after Robert Metcalfe, an Internet pioneer), to be proportional to the square of the number of users, because that is the number of possible directions of communication.[7] This rule does not just apply in the new world of the Internet. Think of the fax machine, urged Lawrence Summers when describing the oddities of the new economy. If there is only one in the universe, it is "a hunk of metal that is best used as a doorstop." But if there are a hundred thousand fax machines, "that is ten billion possible connections." One of the most powerful forces driving communications industries toward concentration is the "network effect." A large network, with millions of customers, has an almost insuperable advantage over a small one that is just starting up.

Many industries in and around the old-established communications business are based on physical networks, such as telephone lines or cable-television systems. In most countries, governments reinforced the natural tendency to concentration that networks always have. They took the view that physical networks that provided important public services were safer in the hands of the state than the private sector; and then typically invented tough laws to make sure that no upstart privateer could compete. Only in the late 1990s did most countries start to privatize these networks and allow competition with them.

When computers came along, they created a new kind of "virtual" network. If you happened to use an Apple Mac, it was hard to send files to your friend who had a PC. Mobile telephones brought new networks too. Europe's GSM (global system for mobile communications) digital standard is a boon, because it covers the continent as well as large tracts of Asia and Africa. The Internet has created lots of new networks, some of them vast. Take America Online's instant-messaging service. American school children use it because their friends use it. Or take eBay. Its dominance of the on-line auction market is another example of a network effect. Sellers use it to reach the greatest number of buyers; buyers use it to reach the greatest number of sellers.

Some of the network effects of communications are entirely abstract. If half of the population watches the Super Bowl or the World Cup Final, the other half pays a certain social cost the following morning when it is unable to join in the gossip around the water cooler. Perhaps the most powerful network of all is language. Anybody who grows up speaking only Welsh or Albanian will be at a tremendous disadvantage compared with those of us fortunate enough to have English as our mother tongue. No wonder that American blockbuster movies such as *Titanic*, benefiting from the combined network effects of movie distribution, social gossip material, and English language, dominate the global market.

The prevalence of systems

One of the main complaints against Microsoft was that it "bundled" some of its products with others, so that customers who acquired a computer equipped with a Windows operating system would also find an Internet Explorer or a Microsoft Network icon on the screen. Another force encouraging monopoly in the communications industry is that its products are rarely used in isolation. Several compatible components are usually designed to work together to form a system. A computer, say, without an operating system; an operating system without applications, such as word-processing programs or spreadsheets; a digital television set without a set-top box to decode the signal; or even the wrong sort of telephone jack for a socket – all are useless.

Systems are so important that companies controlling one part frequently lever their way into others. One example: computer games. Nintendo and Sega, Japanese makers of computer-games machines, also make the cartridges – and influence the content of games. Another: AOL's purchase of Time Warner, with its huge supply of content and giant cable networks. As more and more American homes become connected to high-capacity cable networks, the merged firm will have a powerful hold on the whole system for delivering entertainment, news, and information. A third instance, more notorious: Microsoft's successes in establishing Windows and in dislodging Netscape with its Internet Explorer. These are classic examples of the use of systems to extend market dominance. By "bundling" or incorporating software programs into its operating systems, Microsoft gobbled up much of the market for standalone software for personal computers.

New battles are developing as the Internet becomes available on mobile telephones. In Britain, the SIM cards that carry much of the software for a mobile telephone cannot be bought separately from the telephone. Handsets are sold to be used for one telephone company's services only. In June 2000, Oftel, the government regulator for the telephone industry, investigated BT because it was claimed that Cellnet, its mobile-telephone subsidiary, had programmed some of the handsets it sold to go first to BT's own initial menu page, or portal, Genie.[8] That portal directed users to various shopping and information services. The issue matters, because the sites that telephone users visit generate revenue both as call charges and commission.

. . . . .

The importance of standards

Common standards are essential to networks and systems. They allow users of networks to communicate with one another and make each part of a system compatible with all the others. They are the glue that holds communications together.

Where communications grow fastest, there is invariably a successful standard at work. Think of a fixed-line telephone. Pick up the handset anywhere in the world and it connects (usually) with any one of a billion other receivers, even though the call may have to pass across a dozen telephone networks and finish on a machine in another part of the planet. Or think of the Internet itself, and the way it enables lots of different computers to talk to one another. Or think of the English language, in its role as the world's global tongue. All are hugely successful standards.

Standards bring big benefits for consumers. That much is clear from the irritation caused by situations where none exists. Think of the inconvenience involved in taking a laptop computer abroad with a pocket-full of telephone jacks and a bag of different converters for electric plugs, or the annoyance of having to convert a videocassette recorded by a friend in Germany to play on a television set in the United States.

A widely used technological standard benefits consumers in other ways. It enables manufacturers to enjoy the economies of one large market, for example. Think of how much the rapid development of digital mobile telephony in Europe owed to the early emergence of a common standard, the GSM, which means that the same simple handset can be used to make a call in Ireland or Italy and in Shanghai or Sydney (but not, infuriatingly, in the United States or Japan).

Standards save time. As Bill Gates has pointed out, "Thanks to the common Windows interface, [consumers] can choose from thousands of makes and models of PC, yet will always know how to use the one they opt for."[9] Standards may bring other advantages too. Air-traffic controllers use a standardized set of English words and phrases for their commands to encourage efficiency – and safety.

A successful standard is also self-perpetuating. People select an operating system largely on the number of applications that can be run on it, and people who develop applications do so for the most popular operating system.

In the past, it has been rare for anyone to own a successful standard. But that is changing. In communications, the rapid evolution of new technologies creates lots of opportunities for companies to try to set an industry standard. Microsoft's Windows is the most lucrative example of a privately owned standard. In future, there may be others, such as the standard for the next generation of mobile telephones. The sheer number of standards that networks require raises the probability that some of the most successful will end up in private hands.

When a company owns an established standard, it clearly has a license to print money. That may be a price people are willing to pay in exchange for the efficiency they gain, as long as that price is not too high. As Apple Computer discovered, if a standard is too exclusive, and becomes too great a barrier to entry, it will lose ground to more open standards. One of the greatest strengths of the Internet as a standard is the fact that it is open – anybody can design new ways to use it. Indeed, the ultimate irony about the Internet is that its open standard reduces some barriers to entry into the new economy just as surely as the value of networks raises others.

The dangers of dependence

A dominant standard tends to drive out subsidiary standards. The convenience of standards is so great that many people worry that the market may be badly served if a standard emerges simply because one company gets there first.

One example of the power of a bad standard is analogue television. Of the three main standards – the United States' NTSC (said by cynics to stand for "Never Twice The Same Color"), France's SECAM, and the variations of PAL that are used in most of the rest of Europe – the best is generally thought to be PAL ("Perfection At Last" say wags). The United States, by getting into the market earliest, tied itself to a standard that produces television pictures with inferior definition.

But some of the examples that people tend to think of as proving the perils of becoming locked in turn out, on closer examination, to be unconvincing. In Japan in the 1980s, for instance, Sony's clever little

. . . . .

Betamax cassette recorder fought for the market against Matsushita's clunkier but eventually more versatile VHS machine. An example of the triumph of the "wrong" standard? Probably not. The VHS eventually won the video wars because it offered longer recording time, its price was lower, and its picture quality not substantially inferior. Again, the triumph of Microsoft's DOS over the Macintosh operating system is explicable. Cognoscenti regarded Apple's product as superior, but Macs cost more, and Apple (unlike Microsoft) tended to change operating systems in ways that made earlier software redundant.

Once a standard becomes widely accepted, the costs of changing it may be high. The QWERTY keyboard, universal in the Anglo-Saxon world, saves huge amounts of time because you do not need to learn a different typing technique each time you sit down at a new machine. Many see the QWERTY keyboard as a classic instance of a bad standard locking out a better one: it was supposedly designed to force typists to hit the keys slowly to avoid jamming the old manual typewriters. It has survived into the age of PCs (no danger of key-jamming there), whereas the alternative Dvorak keyboard, designed to allow people to type more quickly and accurately, has never taken off.

In fact, a study of the QWERTY story casts doubt on the supposed inferiority of the keyboard's layout.[10] Its authors, in a subsequent book, pointed out that inferior products do sometimes lock in consumers – but only if the costs of switching are greater than the benefits of using something better.[11] Moreover, the benefits will rise, because inferiority is a spur to come up with alternatives. If Dvorak's merits had really been so great, there would have been a powerful argument for – say – giving away cut-price typewriters or running cheap training courses to give potential switchers encouragement. That is, after all, what is happening with digital television: companies sell the sets at a discount to persuade consumers of the superiority of their new product over the rival analogue standard. A popular standard – even a standard as hugely popular as analogue television – will not survive if a better technology becomes readily available.

Creating Competition

· · · · ·

Given the natural tendency of communications and IT businesses toward concentration and monopoly, it is essential that governments accept responsibility for ensuring competition in the new world of communications, as in the old. The market cannot be relied upon to deliver it. How should governments set about the task?

For all the worries about market dominance in computer operating systems and in the new industries of the Internet, the question is not new. In both the telephone and cable-television industries, the United States wrestled with it for the final two decades of the twentieth century. The issue arose less often elsewhere, simply because the United States was the only country in the world in which both industries were entirely in private hands. In most other countries, the state owned the telecommunications monopoly. Television outside the United States was largely broadcast, not cable, and was generally also either state owned or at least publicly financed. Only in the 1990s, as technology changed, did both industries start to move into the private sector and open to competition around the world. And only then did other governments begin to grasp the problems that creating effective competition involves.

In both industries, two arguments recur particularly frequently. The first concerns the myriad ways in which a big network can shut out smaller rivals to prevent them from reaping the huge economic benefits of connecting with it. A big telephone company may overcharge a smaller one for the right to connect its customers' calls. A large cable-television network may refuse to carry its rivals' programs. These issues arise in newer communications industries, too. In 2000, AOL tried to shut out smaller rivals from its immensely popular instant-messaging system.

The second argument is over the extent to which large, established players should carry obligations that upstart rivals do not. Should big telephone companies be obliged to offer universal service, carrying calls even to places where there is no economic return? Should the main broadcasting network have to run educational or news programs, even if its ratings fall as a result? And should the rules of competition weigh more heavily on the big than on the small?

. . . . .

It may, incidentally, seem obvious that competition should be a powerful force for improving communications. In fact, many governments, in rich and poor countries, took a long time to realize it. When governments owned the telephone monopoly, they found it a handy source of revenue. They fretted about the consequences competition might have for employment, too: state telephone monopolies employed huge numbers of people, who feared for their jobs. In the case of television, they worried about losing control of what their citizens saw on their screens.

Governments that saw the chance of raising even larger sums by privatizing the telephone monopoly worried that, if they threw open the market to competition too quickly, they would devalue their precious asset. Developing countries in particular have been quicker to sell part of their incumbent monopoly than to allow other companies to compete with it. One of the countries that has gone farthest is Mexico, which privatized Telmex, the state monopoly, in 1990. But the sheer importance of this huge company to the Mexican economy and the way it dominated the value of the Mexican stock market meant that it took five more years to begin to open the market to competition. As competition has spread, Mexico has benefited from new kinds of service: prepaid cards for mobile telephones, for instance, account for half the revenues from that sector.

In general, where competition has flourished, so on the whole has the new economy. Where it has struggled, the reverse has generally been true. Creating effective competition in telecommunications is the single most important step that governments around the world need to take to bring the benefits of the death of distance to their people.

The benefits of competition

In communications, as in most industries, competition brings benefits to consumers. In the mid-1990s there was a striking contrast between telephone advertising in the United States – preaching the benefits of talking on the telephone to subscribers who made around three thousand calls per line per year – and advertising in Germany, where subscribers made fewer than one-third as many calls and Deutsche Telekom urged *"Fasse Dich kurz"* ("Keep it short").

In countries that were quick to allow competition in the telephone business, not only do people use the telephone more intensively, but they are also more likely to enjoy new services and state-of-the-art equipment. In Europe it was the mobile-telephone companies that introduced the then-revolutionary concept of billing people only for the time they talked, rather than in blocks of a minute or longer.

The Internet too develops more rapidly in countries where telephone services compete than in those where they do not. A study by the OECD in 1996 found that Internet access across countries with competitive markets was growing six times faster than it was in monopoly markets. Prices were lower, too.

Such issues may seem academic now that most wealthy countries have competitive telecoms markets, yet many poor countries still resist opening their markets. All the evidence suggests, however, that the best way to stimulate the availability of telephones and Internet access in poor countries is to open the field to competition.

In television and radio, competition also multiplies choice. Americans are used to having many television and radio stations, but for most of the rest of the world, choice is a relative novelty. The benefits of choice in television and radio are hard to measure, but the speed with which the audiences of the established public broadcasters declined in the 1990s – in Germany and Spain, for instance – suggests that many viewers and listeners are delighted with their new station options.

Interconnection

Many people assume that, once markets are open, competition will automatically flourish. The experience of the telephone industry in the 1990s shows how false this assumption can be in the communications business, and indicates the continuing role government may have to play to make sure that competition survives.

For instance, America's Telecommunications Act of 1996 allowed a free-for-all in communications. Long-distance carriers could enter local telephone markets, local carriers could offer cable television, cable-systems operators could offer telephony, and so on. Yet the immediate effects were often unexpected. Rapid consolidation was one of them. AT&T, the leading long-distance operator, merged with TCI, the main cable-television company; several former Baby Bells linked up, including Bell Atlantic and Nynex, and SBC and Pacific Telesis; and World-Com, an acquisitive new carrier, bought almost everyone else.

Left to their own devices, telephone giants have all sorts of ways to make life miserable for other telephone companies, including long-distance carriers in the United States and mobile-telephone companies everywhere. They insist that competitors that share a network should pay some part of its cost, for example. Fair enough. But how big a part? Most of the costs of a telephone network are capital. The electrical impulses that new entrants send down the wires do not add to the wear and tear. Besides, a network owner allowed to recover costs – on some definition or other – has no incentive to reduce them. How, too, can the true cost of running the network be decided?

Often, the plight of new competitors is rather like that of a company setting up a new supermarket but using a rival's in-house delivery system. There are endless arguments about how much (to use the supermarket analogy) the newcomer should pay? Whether it can use its own drivers and put its name on the delivery trucks? And how should disputes be settled?

Some of the fiercest fights center on telephone numbers. Customers are reluctant to change telephone companies if it means changing numbers too. Hong Kong's telecoms watchdog once calculated that even tariff discounts of 10 to 15 percent were insufficient to compensate customers for the nuisance and expense of such a change.[12] So who should pay the cost of allowing customers to keep their numbers – the newcomers or the old giants? Many countries have been reluctant to confront such issues.

Similar questions crop up in television. Viewers need special software to receive and decode a signal for digital television, for example, but they are unlikely to want to buy more than one set-top box for the purpose. In Europe, regulators have tried to insist that the boxes be designed so that they can receive signals from several rival companies. The argument is familiar in the United States, where cable-television companies bicker over the terms on which they carry their competitors' programs.

The obligations of size

Theodore Vail, the first chairman of AT&T, built that national giant on the back of the slogan, "One policy, one system, universal service." Once telephone monopolies began to see their market under threat, they too began to talk a great deal about universal service. Public-service broadcasters, such as Britain's BBC, like to point out that they take uniquely seriously the obligation to foster national culture and reflect minority tastes. In a competitive free-for-all, runs the underlying message, these useful things would be lost.

In fact, universal service has proved a relatively manageable problem. Once the obligations involved are clearly spelled out, they become harder to defend. True, the telephone is an essential part of modern life: it can be indispensable for finding a job, reaching a doctor, or staying in touch with family and friends. But exactly the same could be said of the automobile. Yet governments never contemplate providing an automobile for every citizen – and rarely try to fix vehicle prices.

Instead, competition has forced some governments to look more closely at the cost of universal service and at better ways of providing it. Often, the costs of provision have turned out to be quite low. Indeed, new entrants, especially in the mobile-telephone business, have tended to see people without a telephone as a business opportunity rather than a burden. They have invented new products to reach people to whom even governments never promised "universal service." The growth of pre-paid telephone cards, for example, has brought the telephone to millions of people (such as teenagers) whose lack of credit-worthiness would never allow them to have a mobile telephone of their own.

In the case of television, too, competition can reduce the cost of providing public services. If governments want to encourage cultural and educational programming, the wisest way is to put the task out to tender and meet the cost from the same pot as other financial subsidies for the arts or for schooling. America's Federal Communications Commission has tried an alternative approach, obliging broadcasters to provide a certain number of hours of improving television. That tends to lead to long and wasteful arguments over whether a program such as *Rugrats*, say, is educational enough to meet the criteria.

. . . . .

What governments can do

Given the huge economies of scale that big network businesses enjoy, what can governments do? First, they need to open markets to competition, including competition from abroad. That is already under way – on paper. Some seventy countries agreed at the World Trade Organisation to open their markets to foreign telecommunications competition and to set rules for fair global competition, starting in 1998.

The next step is to set up independent regulators – in a surprising number of countries, governments begin by letting the old telephone monopoly decide the basis for connecting with its new competitors – and draw up clear rules. The task of devising rules is made harder by the fact that, in the old communications industries, would-be competitors face two challenges: an existing giant, and economic pressures for concentration.

There are two main approaches. Some countries, such as Germany and Britain, have chosen rules that are deliberately asymmetric, imposing heavier obligations on the old giant than on the new entrants. This can lead to problems. First, the new entrants may not be minnows. In Germany, Deutsche Telekom complains that its competitors are backed by some of Europe's biggest utilities and retailers. Second, it may be hard to know when asymmetry should end. A regulator who starts giving special favors to one group of companies may find it difficult to know when and where to limit them.

The alternative is to split up a dominant giant. America used that approach to reduce the power of AT&T in the 1980s. Now, it seems likely to apply it to a new behemoth, Microsoft. In the case of AT&T, the split helped to create competition in long-distance services and to drive down prices. The question now is, will it bring similar benefits if applied to Microsoft? A split, if badly designed, could stifle innovation and so, in the long run, harm consumers more than it benefits them. That is perhaps even more true for the new communications industries, where change is so rapid, than for the old.

New Media, New Monopolies

· · · · ·

While the telephone and cable-television companies fight rearguard actions to protect their franchises, the new media industries have bred giants of their own: Cisco, AOL/Time Warner, and, of course, Microsoft. Throughout the second half of the 1990s, Microsoft had a succession of skirmishes with the American authorities over the way it exploited the astonishingly powerful position of its Windows software. These culminated in the autumn of 1998 in the filing of a succession of broad antitrust suits by the Department of Justice and the attorneys-general of nineteen American states.

These legal manoeuvres and the trial that followed them in 1999 provided an opportunity to discuss the extent to which new-economy industries were particularly prone to monopoly. One point, argued in his testimony for Microsoft by Richard Schmalensee, one of America's leading industrial economists (and subsequently Dean of the Sloan School of Management at MIT) was that Microsoft was not a monopolist simply because it supplied almost all the market for PC operating systems. Mr Schmalensee (who was retained by the Microsoft side) argued that Microsoft was not doing what monopolists do: maximizing its profits by charging a higher price than it would obtain in a competitive market. Why not? Mainly, he argued, because the software giant had to compete with itself. Operating systems do not wear out, like cars, or get used up, like coffee. So why, having acquired one, would anybody buy another? Why would someone with Windows 95 installed on their PC upgrade to Windows 98? The only incentive is that the benefits of doing so are great enough to justify paying the price. Microsoft's main competitor, in selling a new version of Windows, is the previous versions it has sold.

In fact, the key question was not whether Microsoft was a monopolist but whether it had abused its market dominance, through its ferocious campaign against Netscape's rival Navigator Internet browser and its treatment of competitors and partners. Microsoft lost the case and Judge Thomas Penfield Jackson, who had ruled in April 2000 that Microsoft was an "abusive monopolist," ordered that the company be split. That order will be tested in an appeals process that could last for up to two years.

Is a split the right solution? During the trial, many thought that the best way to tackle Microsoft's operating-systems monopoly, the core of its power, was to create three rival firms, each holding the Windows intellectual property, and a fourth, applications, company. Instead, the courts proposed splitting Microsoft horizontally, into an operating-system company and one producing applications. That sounds, to many, like two monopolies for the price of one, each likely to charge more individually than it would do if it remained a component of one giant firm.

In fact, so the theory runs, each company will have an incentive to innovate at the other's expense. The applications company will have an incentive to create applications for other operating systems, not just Windows, and the operating-systems company to encourage rival applications, undermining the grip of Office. In time, the operating-systems company might start building its own applications, and the applications company might develop its own operating systems. If that happened, two Microsofts would indeed be better than one.

The problem is that dividing a large and powerfully united company is a venture into unknown territory. Some have argued that splits of overweening monopolies in the past brought benefits. The classic instance is the fortune that accrued to John D. Rockefeller after Standard Oil was split into thirty "Baby Oils" in 1911. The babies' market value tripled between 1911 and 1913. However, a study of the break-up found that Rockefeller's stock-market gains were no greater than those of investors in other oil companies at the time.[13]

On the other hand, the trauma might easily do lasting harm without winning proportionate benefits for consumers. Today's threatening monopoly might be tomorrow's sickly giant. IBM, for example, took years to recover from its antitrust case. Microsoft itself faces plenty of challenges. It was slow to grasp the importance of the Internet; its future still seems bound up with the PC, which faces uncertain prospects in a world of networked gadgets; and it will be harder to dominate the open world of the Internet when computing power resides in the network rather than the desktop. In the fast-changing world of cyberspace, launching good new products faster and more cheaply than ever before is the best defense against another company's dominance.

Knowledge and Monopoly

. . . . .

Governments worry plenty about all the characteristics of new-economy industries that predispose them to monopoly. They do not worry enough about the fact that their own policies often reinforce monopolies and discourage innovation.

The value of many new-economy industries lies, not in the physical objects they produce, but in the knowledge embodied in them. Indeed, a successful Web site, such as Alta Vista, has no physical presence, in the way a successful factory or store does. It is a pure knowledge business.

Now knowledge, as Thomas Jefferson famously noticed, has a wondrous property. "If nature has made any one thing less susceptible than all others of exclusive property, it is the action of the thinking power called an idea," he wrote. "No one possesses the less, because every other possesses the whole of it. He who receives an idea from me, receives instruction himself without lessening mine; as he who lights his taper at mine, receives light without darkening me."

But that very characteristic of knowledge – the ease with which it can be shared – creates a problem for knowledge industries, and for regulators. For governments have to decide how far to protect the monopolies of the knowledge industries and how far to lean against them. Protect? Certainly: patents have become the main legal tools for guarding the products of the knowledge industry from competition, and they are in the gift of governments. So governments need to consider whether tolerating the new monopolies is likely to stimulate the development of world-beating knowledge products.

The growth of patents

Patents are the strongest title to intellectual property. They give holders a claim over ideas encapsulated in a work and not just (as with copyright) over the particular form the work takes. They start from a perfectly respectable aim: to reward inventors, and thus encourage future invention. They achieve that aim by granting the patent-holder (in the United States) a legal monopoly for twenty years from the date of

· · · · ·

application. But it is extremely difficult to strike the right balance between encouraging inventors and stifling new products.

Led by the United States and Japan, the number of patents being granted is rising steeply. [Fig 10-2] This is partly for good reasons. The value of knowledge has grown relative to other assets, and so there is more knowledge for companies to clamor to patent. Much the same happened at the end of the nineteenth century, another period of rapid innovation, when Thomas Edison set the record (which he still holds) by acquiring a staggering 1,093 patents[14]. Edison knew better than anyone that, while there might be no loss to an inventor if somebody else thought up the same idea, there would certainly be a substantial financial loss if a rival embarked on its commercial exploitation.

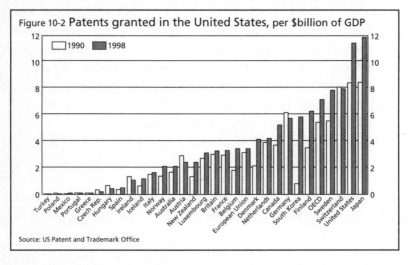

Figure 10-2 **Patents granted in the United States,** per $billion of GDP

Source: US Patent and Trademark Office

However, new kinds of patent are now being issued, for computer software and even business methods. Among the ideas to which America's patent office has given monopoly protection are such simple ones as group buying, the matching of professionals with other people seeking their advice, and one-click shopping.[15] "Software and algorithms used to be unpatentable," complained James Gleick in an article in the *New York Times*. "Recent court decisions and patent-office rule-making have made software the fastest-growing category, and companies are rushing to patent the most basic methods of doing business."[16] No wonder that every self-respecting e-commerce company went out and hired a tame patent lawyer.

A classic example is that of Dell Computer, which in the mid-1990s lodged several applications to patent, not its PCs, but its new way of doing business by building machines in response to orders. By 2000 it had seventy-seven patents protecting different parts of the complex building and testing process involved in its build-to-order system. Henry Garana, vice president of intellectual property at Dell, confesses that, at the time, many people regarded these patents as a waste of effort. Now, he says, "They make people go away. In this business, that's what matters."[17]

Companies initially saw such patents as Mr Garana did, as a way of defending a good idea – or even a business. Both Texas Instruments and National Semiconductor avoided bankruptcy in the early 1990s by the aggressive use of patents.[18] Gradually, companies realized the commercial value of patents as a form of property and became more systematic in their patenting policies.

IBM, one such company, now takes out ten patents every working day, and claims to have generated $30 billion in revenue from licensing the use of its patented technologies.[19] Financial-services firms and accountants are starting to patent their techniques (a classic instance is the patented Merrill Lynch Cash Management Account), something that in the past would never have occurred to them. Other companies have sprung up to make their livings entirely from patents. One such is Walker Digital, which has patented the idea of reverse auctions, in which customers set a price they are willing to pay for something, such as an airline ticket, and then airlines decide whether to meet it. The idea gave birth to a high-profile company, Priceline.[20]

Allowing a company to patent the concept of the Dutch auction, an idea as old as the Dutch themselves, may seem bizarre. Even more bizarre is that companies should patent ideas that require technologies that do not yet exist. If a company can plausibly describe how something might be done, then it can patent it. Telecom Partners, a company that builds patent portfolios, has created a sheaf of portfolios on ways of targeting television and video advertising over the Internet. The most valuable patents concern ways to send different advertisements to different people through cable and high-speed telephone lines, a technology that is barely formed.

The patent office argues that such ideas fulfil the rule that ideas embodied in a patent application must be novel and useful, and must

. . . . .

not be obvious. But it is hard to see how the creation of such broad monopolies can fail to obstruct innovation by others. Not surprisingly, a growing number of court cases are testing the law. In 1999, Amazon got a patent on one-click shopping and promptly won an injunction against Barnes & Noble, claiming that the latter's Express Lane infringed the patent.

The need for new rules

Clearly, inventors need some protection for their ideas. But Thomas Jefferson (himself a keen inventor) would have been aghast to see how much protection patents now provide. Their reach has become too wide and too long to be appropriate for the new world of knowledge. "This is a disaster," says Lawrence Lessig, a Harvard professor and expert on Internet law. "In my view, it is the greatest single threat to innovation in cyberspace."[21]

Certainly the most popular argument for strong patent protection, that it is a precondition for innovation, is decidedly questionable. An important study of industries such as software, semiconductors, and computers finds that a strengthening of patent protection in the 1980s was not accompanied by a surge in research and development intensity and productivity. (Microsoft, founded in 1975, received its first patent in 1986 – for a sort of hinged box.)[22] Instead, R&D stagnated and even declined in those industries and firms that patented most. The problem, argue the authors of the study, is that innovation is sequential. Each innovation builds on a previous one, as Windows built on DOS. It is also complementary. Each potential innovator takes a different line of research, increasing the probability of success – as in the case of the many different commercial approaches to voice-recognition software. Patents are an impediment to both processes.[23]

Past waves of invention and consequent patenting have produced similar tendencies to overreaction. In 1882, when ruling on a case involving boat-propeller technology, America's Supreme Court complained that "An indiscriminate creation of exclusive privileges tends rather to obstruct than to stimulate invention. It creates a class of speculative schemers who make it their business to watch the advancing wave of improvement, and to gather its foam in the form of patented monopolies, which enable them to lay a heavy tax on the industry of the country, without contributing anything to the real advancement of the arts."[24]

One possible reform might be for the United States to consider Europe's approach to patents, which has been more limited. In particular, after a patent has been issued, the competition has a statutory right to oppose it, and is often successful. The result is arguably to discourage bad patents.

In addition, patent protection lasts too long for the fast-moving innovators of the Internet. Even Jeff Bezos, founder of Amazon (which has benefited handsomely from patenting its "one-click" purchasing method), is not convinced. He has suggested in an open letter posted on the Internet that patents for software and business methods should have a shorter life than other patents – three to five years, say. Even that is a long time in Internet years. Governments need to grant fewer patents, with shorter lives. Otherwise, far from protecting innovation, patents are likely to become a serious barrier to new entrants with better ways of doing things.

An Open Mind, An Open Market

. . . . .

The industries of the new economy are driven, even more than are those in the old economy, by the desire to lock up markets and leverage dominance from one product to another. The instinct may be stronger, but it is hardly new. Neither is the need for antitrust vigilance, although regulators are almost always better to err on the side of doing too little, rather than risk doing too much, when technology is changing fast.

Indeed, the power of technological change is so great that there is more danger in regulating too early than in leaving things too long. The right approach in every part of the communications and IT industries should be the least regulation that is consistent with lively competition. For many countries, the fastest way to foster competition will be to strip away controls and rules that deter new entrants. The more competition emanates from unexpected directions, crossing conventional industry divides and leaping national borders, the more innovation consumers will be likely to see.

11

Society, Culture, and the Individual

A lunch at the World Economic Forum in Davos, January 1999, for people with an interest in telecommunications in North Africa and the Middle East, one of the more underconnected parts of the industrializing world. A Western technophile enthusiastically expounds, to a predictably male audience from the region, the economic benefits that the communications revolution can bring to developing countries. The response is subdued. Afterwards, another Westerner, a woman with long experience in the region, leans across to the speaker and murmurs, "Of course, what many of these people really worry about is the effect of home access to the Internet on their wives. It will give them all a window on the world."[1]

Like all good revolutions, the technological changes that are rushing forward are essentially liberalizing. They bring to the mass of people what was once available only to the elite, and to the individual what was once available only communally. Other technological changes have had similar effects: think how the radio enabled every family to hear Yehudi Menuhin play a violin concerto, and how the Sony Walkman then allowed every individual a personal performance. Both the mobile telephone and the Internet are personal technologies, giving individuals access to services (a telephone call, a research library) that were once communal – or even inaccessible.

Judging the long-term social impacts of new technologies is extraordinarily difficult. As Arthur C. Clarke once pointed out, people exaggerate the short-run impacts of technological change and underestimate the long-run effects. Really big technological changes permeate our homes, our personal relationships, our daily habits, the way we think

. . . . .

and speak. Consider the links between the automobile and crime, or between electricity and the skyscraper, or between television and social life. Each technological advance had consequences that nobody could have foreseen when they were new. The revolution in communications will have results that are just as pervasive, intimate, and surprising.

Already, there are unexpected glimpses of new ways of doing things. Young socialites in London, each toting a mobile telephone, claim that it has already become possible to squeeze more fun into a single evening. Life is becoming more fluid, more immediate, with lots of scope for sudden changes of direction. (Soon a text message on a mobile phone might tell people: "Let's not meet at the café, let's go straight to the movie – and I'm copying this message to everyone else in case they want to join us after for a drink. I'll leave the phone switched on so that they can see where we all are.") It allows what might be dubbed "real-time living."

Despite such cheerful uses of new technology, many people fear its social effects. They see a society of isolated people, stuck indoors, glued to a screen, losing the taste for real human contact and experience. They worry about the exclusion of the poor, the old, and those too inept to learn how to configure their modems. They imagine a new class of technological have-nots, as socially deprived as anybody without an automobile or a driving permit is often presumed to be. They see an Orwellian world of lost privacy. In countries other than the United States, people fear a future in which everybody speaks English and thinks like an American, with cultural diversity engulfed in a tidal wave of crass Hollywood values.

In fact, the main impact of the death of distance is to make communication and access to information in all its forms more convenient. On balance, that will surely be good for societies everywhere, although the nature of the effect will depend on why people communicate and on what knowledge they choose to acquire and how they use it. Like the automobile, telecommunications technology is a tool that can be used for bad purposes as well as good; but it should, overall, make people's lives easier and richer. More communication is generally better than less.[2]

In broad terms, the societies of the rich world are being altered in at least four main ways. The role of the home is changing, as it reacquires functions it lost in the last century. It is becoming not just a workshop, but a place where people receive more of their education, training, and health care. New kinds of community are emerging, bonded electronically across distance, sharing work, domestic interests, and cultural backgrounds. The use of English is being enhanced, as it becomes the global tongue (although many regional languages will revive as well). And, finally, the young, especially the intelligent, well educated young, will benefit, and young countries will gain a potential advantage over older ones.

So society is changing. Communities will take different forms and people will take for granted things that now seem strange. But the biggest changes will take the longest. The death of distance will shape the world of the mid twenty-first century more profoundly than it will the world of the next decade.

Work and Home

· · · · ·

The falling price of communications will affect where people work and live. The demarcation between work and home will blur. In its place will be a shift in the location of work, a new role for cities, and a new role for the home. Of these, the third may be the most profound. The home will once again become, as it was until the Industrial Revolution, the center for many aspects of human life rather than a dormitory and a place to spend the weekend.

The location of work

For many folk, work is becoming not so much a place as an activity: something you do, rather than somewhere you go to. A primary shift in the location of work means that fewer people stay in conventional offices for their entire working week. Many already spend some time "telecommuting" from home. Others increasingly work on the move, or in places that never used to be thought of as offices: hotel rooms, cars, airport departure lounges. Some mobile workers, including sales representatives, service personnel, and repair crews, are engaged in various kinds of maintenance or in delivery (of people as well as goods). Many others – including some well paid folk such as non-executive directors, consultants, and one-person agencies – are "portfolio" workers, undertaking projects for several clients at once, combining a series of freelance jobs under contracts with different companies.

At the same time, offices are beginning to serve two differentiated roles. Some are effectively factories – call centers, data mines, back offices. Others are becoming more like clubs.

Working from home

When the communications revolution first gathered speed, some people argued that it would lead to more people working from home. That does indeed seem to be happening. In 1999, almost one in ten American adults said they telecommuted from home.[3] But the proportion may rise more slowly than the proportion who work on the move.

Several experiments have suggested that home-based teleworking, properly organized, can bring benefits such as greater job satisfaction and less stress. An experiment conducted in the far north of Scotland by British Telecom, a large British telephone company, found that home-based directory assistance operators, connected to a supervisor and the customer by a telephone link and a computer link, were particularly reliable.[4]

For the moment, companies see telecommuting more as a perk than as a way of boosting productivity. They often use it to retain a good employee who has been threatening to leave because of difficulty in balancing the demands of work and family, or whose spouse is moving to another part of the country. Some companies have a more subtle approach. Ford, for example, has a plan to give eligible employees a computer, printer, and Internet access at home for a nominal fee. The aim, the company says, is to encourage employees to develop technology skills at home, improve access to corporate information, and foster new ideas.

However, electronic home-working has some significant drawbacks. If equipment goes wrong in a worker's home, the cost of repairing it might be higher than would be the case if it were in a central office. As long as their equipment is out of action, so too are the teleworkers. And then there is the problem of isolation. In the mid-1990s TBWA

Chiat/Day, an advertising agency, found that its offices in Venice, California were too small for its fast-growing work force. It gave everyone on its staff a laptop, a locker, and a cellphone and told them to come into the office only when they needed to. The experiment was a disaster – employees grumbled that they lacked creative interaction. In 1998 the company gave in and moved into offices large enough for all the staff to have their own desks and plenty of open spaces for all that creative interacting.

Along with worries about isolation go, perversely, worries about excessive contact. Employers can reach their employees wherever they are. New understandings will be needed in future that limit the intrusion into workers' lives. Improvements in communications have always threatened privacy. In the 1850s James de Rothschild, one of the illustrious banking family, grumbled that the coming of the telegraph meant that "even when he went to take the waters from his summer holiday, there was no respite from the business: 'One has too much to think about when bathing, which is not good.'"[5] The Internet, the laptop, and the mobile telephone have merely democratized that intrusiveness.

And governments may also worry about how to extend to white-collar telecommuting workers the rules that protect their office colleagues and, indeed, blue-collar homeworkers. In November 1999, America's Occupational Health and Safety Administration suggested that employers might be responsible for "correcting hazards" in a telecommuting employee's home. In the face of protests, the OSHA later softened its line. But the debate raised a real issue: working at home sets employees free, but also avoids some of the onerous obligations on employers.

Workers in transit

Changing communications have made it easier for companies to use mobile workers efficiently. A controller can use satellite-positioning equipment to track a repair crew, a computer to calculate the most efficient routes for workers to take, and mobile telephones to maintain contact between despatchers and workers. Thus maintenance folk at Centrica, a British domestic-repair company, no longer gather at the local depot each morning to pick up spares and a worksheet for the day. Instead, they work from home, and log on to a laptop with a wireless link to take jobs as they come into the firm's call centers. The call center feeds customer details through a computer that allocates jobs on the basis of which engineer is nearest and what skills are required.[6]

Such travelling teams pose special management problems. Companies sometimes find it difficult to instil a sense of corporate culture into workers they rarely see. What will be the glue that holds a company's dispersed road warriors together? Perhaps there is a new role here for secretaries and personal assistants to update travelling staff on office gossip, or an extra task for the corporate communications department, which will need to spend as much time communicating the company's culture and goals to staff as to the outside world. Regular face-to-face meetings and effective communications will be essential to good management of workers-on-wheels.

Ironically, better communications will mean that many senior employees, too, find themselves spending more time on the road and working almost anywhere but in the office. To hold together a business with global scope, which may have production, distribution, and alliance partners spread among several countries, requires not only an efficient electronic network but an enormous amount of travel by managers and suppliers. For these people, living near a good airport may be even more important than living near the head office. They will need an "office in a box": portable communications that connect – reliably and ubiquitously – with the office database and the Internet.

These workers may also operate from restaurants, at conferences, in golf clubs. Such places provide a more congenial backdrop than the office for the face-to-face contacts that are central to industries whose products embody mainly knowledge and ideas. A senior partner in an advertising agency cannot deal with clients entirely by telephone, any more than can a high-powered lawyer or a Hollywood agent. These personal contacts require a structured informality, a blend of the personal and the professional. The telephone and electronic mail may reinforce basic contacts, but they are no substitute for face-to-face encounters.[7]

The future of the office

While work is intruding more into home life, home life intrudes more at work. On some estimates, almost one-third of the time American workers sit at their desks is now spent on something other than the boss's business. So 90 percent of American workers admit to surfing recreational Web sites during office hours, 84 percent say they send personal e-mails from work, and more than half do a spot of cybershopping in company time.[8]

When workers are in their offices, they are likely to be in one of two quite different places: white-collar factories, or clubs. Call centers and data warehouses, for instance, have many of the characteristics of a factory. They can be situated far from customers (indeed, on a different continent); they involve routine work, sometimes around the clock; and they employ mainly people with moderate skills, doing repetitive tasks that are usually relatively easy to learn.

Other offices are coming to resemble clubs, a transformation first spotted by Charles Handy, a British management guru. They are places for the social aspects of work, such as networking, lunch, brainstorming, and catching up on office politics. The office-as-club aims to foster a sense of fellowship (or corporate culture) among employees, stimulating conversation and bonding.[9] One pioneer of such offices is Accenture, formerly Andersen Consulting, which has slimmed down its expensive city-center offices in Paris and Boston, arguing that its best people spend most of the their time on the road or in clients' offices. Instead, staff can book into a "virtual private office" for the day, in buildings that have lots of "huddle rooms" for brainstorming and open spaces for encouraging coffee and conversations with colleagues.[10]

In the future, then, more work will be done outside offices. Some will take place in the home, some in transit. That will have long-term consequences for cities, where work has increasingly been located.

The future of cities

In half a century's time, it may seem extraordinary that millions of people once trooped from one building (their home) to another (their office) each morning, only to reverse the process each evening. In the past century, people have tended to live farther and farther from their place of employment. But commuting wastes time and building capacity. Roads and railways must accommodate the weight of rush-hour traffic. One building – the home – often stands empty all day; another – the office (often in the most expensive part of town) – usually stands empty all night. In some cities, almost all the time saved by the decline in working hours in the twentieth century was gobbled up by increased commuting time.

Now commuting is becoming common in industrializing countries too. The rule of thumb seems to be that, in cities of one million people, workers must travel an average of three miles to their jobs; in cities of five million, they travel seven miles.[11] One of the main reasons for this trend has been the concentration of employment in large units. Through the first half of the twentieth century, the factory, and then the office, tended to grow. In the case of factories in most rich countries, this trend had begun to reverse by the 1970s, mainly as a result of the rise in productivity. Now this reversal will gather pace, permitting work to become more dispersed. The dispersal of work will alter the nature of our cities and towns, reaffirming their importance as places where people live and as centers of entertainment and culture.

Some cities that were hollowed out in the second half of the twentieth century, as work moved to the perimeter and habitations even further out, are starting to attract people back to live in the center. This is to some extent an effect of demography: families may want the space of the suburbs, but young singles and couples and the growing number of ageing baby-boomers whose children have left home are more willing to trade space for bustle.

Cities with attractive architecture and plenty of shops and cafés providing lively street life will delight those who relish the interest and anonymity of living in a crowd. Cities will become safer, too, thanks to the wider use of electronic surveillance. Cities already thrive as centers of entertainment and culture: places to which people travel to stay in a hotel, visit a museum or gallery, enjoy a restaurant meal, or hear a concert or band. Many kinds of work require entertainment. Managers increasingly want to take valued customers to a theater, a restaurant, a party. So some of these delights will be paid for, directly or indirectly, by the companies that use their offices as clubs.

Suburbs and rural towns will benefit from the telecommunications revolution in other ways. Many people may once again "live over the shop" and work in their communities, rather than commuting. The benefits will include more revenue for local stores and other services, as workers stay close to their home towns during the week; more opportunities for delivery services, since ordering over the Internet or telephone is easier if someone is at home when the product reaches the doorstep; and less local crime, because homes will be occupied and streets will be busy during the day, making them safer for everyone.

All of these changes have a down side. The blending of leisure and work often means, in practice, that work intrudes into leisure: it makes the more forceful demands. One study found that a quarter of Americans with Internet access at home used it to work more hours at home – without a corresponding cut in the hours worked at the office.[12] Homebound or travelling workers may yearn for the companionship and comfort of the conventional office. And, of course, not every downtown office complex, if deprived of office tenants, will be quickly transformed into a glittering entertainment center. But on balance, the direction of change may be to restore communities, to improve the quality of cities, and to give people more control over their working lives.

The future of homes

Among the most striking changes brought about by the death of distance is a shift in the role of the home. People not only entertain, relax, and sleep at home; they increasingly find there a range of services, from health care and education to investment and employment. From their homes, people can study any subject from astronomy to zoology, seek legal advice, participate in a political debate, or bid in an auction.

Home-based services to come will include monitoring, and sometimes repair, of domestic machinery without the need for a home visit; supervision of the housebound sick or elderly; and a level of security observation currently confined largely to offices and the wealthy. Some caregivers already install a camera to enable parents at work to watch their children at play. In time, some labor-intensive monitoring services may be provided by countries with cheap labor, with a hotline back to the neighborhood so that on-the-spot help can be available quickly.

Such changes will alter the design of the home. Architects have been slow to catch up with its mutation from a place where people consume (whether meals or entertainment) back into a place where people also produce (today's homes may be equipped with as much computer power as a large factory had in the 1970s). They now need to devise better ways to accommodate the home office.

For many people, the home office remains a problem: it doubles as a spare bedroom or crouches uneasily in a corner of the living room. Until the home office finds a permanent resting place, home design will need to be adaptable, so that flexible use of space can match flexible work patterns. The loft, with its vast open space, may be a more useful prototype for the twenty-first century home than the houses of the 1950s and 1960s, with their small, purpose-designed rooms.

New Communities

The concept of cyberspace – a computer-generated, multidimensional world in which people live in virtual reality – was conceived by William Gibson, a science-fiction writer, in a book called *Neuromancer*, published in 1984.[13] The disembodied world he presciently portrayed symbolizes for many people the danger that electronic communications will be isolating and inhuman, as people find a social life in the chat rooms of the Internet.

But it is perverse to imagine that a technology that makes it easier to communicate should simultaneously reduce human contact. New forms of communication may change the nature of that contact – as the telephone call has to some extent replaced gossip in the village street – but, more probably, they also increase the variety of ways people can and do communicate. At times, people prefer the privacy of a telephone call to a gossip in full public view.

The main impact of better and less expensive communications is likely to be to create new ways to socialize and build communities of interest, independent of geography. Both will enrich people's lives and mitigate the effects of separation that go with the increase in international migration, overseas employment, and business travel.

The future of socializing

Communications alter social contact. By far the most popular use of the Internet is electronic mail, and social calling constitutes the fastest growing area of international telephone traffic.

Some people will want to talk to friends, some to strangers. In terms of time spent on the line, talking to friends is the telephone's main use. Such social calls tend to last for twenty to thirty minutes – far longer than the brisk two to three minutes typical of a business call – and to be made for the most part by women. One study found that the predominant overseas caller from Australia was a young professional woman, born in another country, making long social calls to friends back home.[14]

International migration, business travel, and tourism all increase the number of people separated from their friends by distance. More telephone calls go from Germany to Turkey, for example, than from Germany to the United States, because although the United States is a more important trading partner for Germany than is Turkey, Turkey has been a bigger source of immigrants. Electronic communications enable friendships and families to survive separation.

In the days before inexpensive telephone calls, people wrote to each other more frequently. That had some advantages. The letters of John Keats or Queen Victoria are preserved for posterity as telephone conversations are not. It also had drawbacks. A single minute's telephone conversation contains around two hundred words, or perhaps as many as a handwritten letter might convey. Talking thus saves time. Now the Internet may partly replace the telephone call. Indeed, when people have access to the Internet at home, some of the time they spend on-line would once have been spent talking on the telephone or seeing friends. A large study conducted by Stanford University found that the main reallocation of time by regular users is from watching television (on which 60 percent of users apparently spend less time) and reading newspapers (34 percent). However, 25 percent of regular users divert time from telephoning friends and family, and 13 percent from seeing them.[15]

A loss? Not necessarily. The Internet offers forms of communication that some people may prefer. It facilitates contact with strangers: a typed message need carry no information about personal appearance, age, race, or gender. Bulletin boards and chat rooms offer scope for one-to-many contacts that create opportunities for discussion and debate.

Indeed, as it becomes easier and less expensive for individuals to communicate with others electronically, people choose to chat for longer, whether on the telephone or on the Internet. It is not published content – news reports, sports scores, reference material, and so on – that people most crave on-line. The most compelling content of the electronic world is simply other people. The pleasures of communicating beat even the output of Hollywood, and eat into the time people spend watching packaged electronic entertainment. Social life is thus enriched, not impoverished.

Virtual communities

Chatting to strangers becomes easier when based on a shared interest. One of the earliest uses of the Internet was to connect geographically scattered groups with shared minority interests. These "virtual communities"[16] of people linked electronically may meet occasionally for what cyber-enthusiasts call "face time" – or may never meet. Such "horizontal" communities, scattered around the world, may have more in common with one another than with their next-door neighbors.

Electronic communities of interest have long existed in academic life, medicine, and science, where so much of the Internet's early development occurred. Professors of Anglo-Saxon, neurosurgeons, and earthquake specialists may all be rare beasts, working on their own or with others in equally abstruse fields. But they usually know their colleagues around the world from their publications, from conferences, and now from the Internet.

Such horizontal contact becomes more important in commercial life as companies fragment. It might allow the emergence of on-line equivalents of the professional and craft guilds that united workers in the Middle Ages, for example. If growing numbers of employees switch to

self-employment, they may seek out on-line associations to set standards for their trade, help them sell their skills, and provide professional and social support.[17]

In politics, on-line communities can be used to help to build cross-
border alliances that challenge specific organizations or their actions.
Environmentalists in many countries, for example, might unite in battle against what they perceive as a destructive proposal in one country
that is financed by a donor or bank in another. Jody Williams, a cam-
paigner against landmines, organized a thousand human-rights groups
on six continents in a fight that eventually led to a near-universal ban
on landmines and won her the Nobel peace prize. When journalists
asked how she did it, she replied simply, "E-mail."[18]

Bulletin boards and chat rooms have become increasingly specialized, catering to every taste. Such contacts encourage self-help groups
to flourish. They make it easier for people to draw on the collective
experience of other individuals, rather than resorting to professionals
for advice on matters from medication to child-rearing. When, to take
just one example, parents want to debate the pros and cons of spanking naughty children, some of them do so on Web sites such as
iVillage.com and *Oxygen.com*.[19]

The Internet may even stimulate community involvement, thanks to
sites such as volunteermatch.org. This enterprise, launched in 1997 and
based in Palo Alto, lists thousands of opportunities to provide voluntary
service, placed by American non-profit organizations. Type in a zip
code, add the kind of work you want to do, and a raft of opportunities
appears on the screen, including some for "virtual volunteering" which
enable people to do good from their own desks by offering services such
as free programming or graphic design. By early 2000, the site's search
engine had found voluntary work for some eighty thousand people, an
amazing achievement given the difficulty charities often have in finding helpers.[20]

Electronic cultural and ethnic communities help to strengthen ties
that distance might otherwise fray. The Internet provides a meeting
place for people separated from their native lands. Scotland's clans,
those strange and anachronistic Highland tribes, received a new lease
of life from Mel Gibson's film *Braveheart* that has been perpetuated by
the Internet. North American Scots can find electronic listings for local
Highland Games (displays of ancient sports such as putting the shot

and tossing the caber) or search for the appropriate tartan at sites such as greatscotshop.com. They can even buy a kilt on-line, with all the proper paraphernalia to accompany it.

All sorts of other groups have learned to find one another on-line. A fascinating study of the use of the Internet by Trinidadians shows it has strengthened both the nuclear and the extended family among the scattered citizens of an island with a long history of emigration. Some of the country's churches use it to bind together congregations that may be dispersed across continents. And the large Trinidadian diaspora, now bound together by many Web sites, seems to have fostered a growing sense of national identity and culture.[21]

One of the most basic building blocks of cultural identity, a knowledge of one's origins, has been transformed by electronic communications, and especially by the Internet. To research a family tree once meant a letter or visit to a registry office, perhaps in another country. Now, genealogy has become one of the main recreational uses to which individuals put the Web. When, in 1999, the Mormon church opened Familysearch.com, the world's largest genealogical Web site, it received 230 million hits in its first three days.[22]

Inexpensive international telephone calls combined with the Internet allow exiled people to continue to practise their native language. Homesick emigrants can increasingly shop from a distance for delicacies from back home, watch television programs originating from their home countries, read their home newspapers, and follow their home sports teams.

In these ways, electronic communications will reinforce cultures that might otherwise be damaged by distance. Not everyone will welcome that. Some of the most successful Indian and Pakistani Web sites, for example, enable emigrants from those countries to find a spouse from their home community. That may slow the process of integration. In Bradford, a town in the north of England with one of the country's highest concentrations of a single ethnic group, many of the Muslim Mirpuris from Pakistani Kashmir who first came to work in the textile business in the 1950s have still not integrated. One reason is a tradition of marrying children to spouses from back home: 58 percent of the community's marriages are to partners from Pakistan. Local white community leaders fear that an effect of the Internet will be to prolong the divide.[23]

Language and Culture

. . . . .

The stock-in-trade of electronic media will be language and ideas. With the death of distance, many countries fear the power of American culture and the English language. They worry that their own languages will be swamped and their cultures and traditional industries overwhelmed. Both fears are largely unfounded.

Electronic media affect language in three main ways. They alter the way language is used, they create a need for a global language that will most likely be filled by English, and they influence the future of other languages. In that last case, one of the main impacts of new communications will be to lower the entry barriers to cultural industries such as television and movie-making.

A new linguistic style

Electronic communications have been changing the use of language for over a century. To the delicate social question of how to answer the telephone, Alexander Graham Bell came up with the following solution: say "Hello." Developments in telecommunications have brought other cultural innovations. The telephone, for instance, made something that had been a rarity – a conversation with an unseen person – into a common experience. The telephone-answering machine or voicemail produced new versions of the monologue (which, to judge by the messages most people leave, few folk enjoy delivering). And radio sports commentary created what one American linguist describes as "a monologue...directed at an unknown, unseen, heterogeneous mass audience who voluntarily choose to listen, do not see the activity being reported, and provide no feedback to the speaker."[24]

Electronic mail and Internet chat have produced yet another linguistic innovation: the written conversation. Unlike a reply to a letter, which takes at least forty-eight hours to arrive (although in Victorian England responses were much faster), a reply to an e-mail can be more or less instantaneous. And e-mail has encouraged a vast number of people, many of whom may hardly have written a personal letter in

. . . . .

their lives, to correspond. The revival of writing, even if only electronically, is surely a cultural trend to be welcomed.

All that remains is to import into e-mail some of the rules that letter-writers once took for granted, such as the need to avoid writing in anger – or at least to save a furious missive and reread it the next morning before despatching it. Michael Eisner, chairman of Walt Disney, argues that e-mail has increased the intensity of emotion within his company and become the principal cause of work place warfare. "With e-mail," he notes, "our impulse is not to file and save, but to click and send. Our errors are often compounded by adding other recipients to the 'cc' list and, even worse, the 'bcc' list. I have come to believe that, if anything will bring about the downfall of a company or maybe a country, it is blind copies of e-mails that should never have been sent."[25]

The coming global tongue

For the electronic media to work efficiently as global carriers of language, they need a common language standard. Standardizing makes communication easier and cheaper, creating a virtuous circle so that more of it occurs. English has emerged as the necessary standard: the default language, as it were, or the linguistic equivalent of Windows or the GSM mobile-telephone standard. Used by something like a quarter of the world's population, it makes many forms of communication, including trade and foreign investment, less expensive than they would otherwise be. In this, it is like a shared currency or a shared set of tastes. It has the advantage of all networks: the more people can speak it, the more everybody gains as a result.

The dominance of English is unprecedented. It is spoken by more people as a second language than as a first. No previous language has been in such a position. Three-quarters of the world's mail is written in English; perhaps a billion people – one human being in six – are learning it.[26]

Until now, the spread of English has depended on two factors: the legacy of colonialism and the emergence of the United States as the world's largest commercial power. Most of the new countries that have emerged in the past half century have given English a special role, making it the dominant or official language in more than sixty countries.[27]

Otto von Bismarck, Germany's famous chancellor, foresaw the American commercial hegemony more than a century ago. Asked by a journalist in 1898 what he thought was the decisive factor in modern history, he replied "The fact that North Americans speak English."[28]

In the future, the spread of English will be driven by two additional factors. The United States is easily the world's largest net exporter of intellectual property (Britain is the second largest), and English is overwhelmingly the main language of the Internet. People who buy intellectual property, whether Madonna or Microsoft, often buy English as part of the package. About 80 percent of all Internet sites are in English. [Fig 11-1] With some subjects, the proportion is higher. A study carried out in 1996 found that almost all the scientific material on the Internet was in English.[29] There is a good reason for the commercial dominance of the language: a German company, wanting to reach customers in France, Sweden, and Greece, can do so most simply by putting its material on-line in English rather than in German.

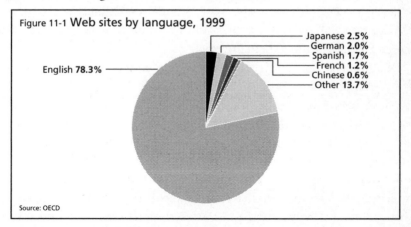

Figure 11-1 **Web sites by language, 1999**

English **78.3%**

Japanese **2.5%**
German **2.0%**
Spanish **1.7%**
French **1.2%**
Chinese **0.6%**
Other **13.7%**

Source: OECD

The dominance of English on the Internet follows inevitably from the dominance of Americans among Internet hosts and users. As the proportion of non-native English speakers using the Internet has risen, other languages have become more widely used. But English will probably remain disproportionately important on the Internet, creating a large new category of English users: those who can write the language colloquially but cannot necessarily speak it.

What might be the alternative to English as the electronic lingua franca? Conceivably, another dominant language might emerge (Chinese or Spanish, perhaps). Or advances might be made in machine

. . . . .

translation, already widely available on the Internet. For the moment, though, machine translation leaves much to be desired (see, for example, Figure 11-2).

As world trade in on-line services grows, the prominent role of English will initially give those who speak the language an advantage over those who do not. India and Jamaica have built their data-processing industries partly on the use of English. The greatest advantage of all will accrue to those of us lucky enough to grow up speaking English. As English becomes a truly global standard for communicating, that advantage will fade. Already Holland, with probably the world's highest proportion of non-native English speakers, has a large and growing English-language publishing business.

Other languages

If English is established as the world standard language, what are the implications for other tongues? Cultural homogeneity? In fact, and paradoxically, they may benefit too.

Digital television, with its multitude of channels, will make it easier and less expensive to produce and distribute niche programs in minority languages. The Internet may also protect subsidiary languages, for two reasons. The limitless capacity of cyberspace means that languages do not compete with one another head-on, as they have done in the past. A Danish rock festival can advertise on-line in English, German, and Swedish – but it can also market itself in Danish. In addition, the Internet offers an inexpensive way for speakers of exotic languages to keep in touch with one another. One description of the World Wide Web found "discussion groups in more than sixty languages, at which point I stopped counting." This explorer discovered people conversing in Aragonese, Armenian, Basque, Breton, Cambodian, Catalan, Esperanto, Estonian, Galician, Hmong, Macedonian, Swahili, Welsh, Yoruba – and so on and on.[30]

Indeed, one of the clearest lessons of the new technology is that it increases the scope for niche communities of all sorts. The young have always tried to segregate themselves from their elders by developing their own incomprehensible slang. Armed with mobile telephones, youngsters in Britain and Japan have done exactly that. Sending short

Figure 11-2 **Translation aids communication**

Hans Schultz
Gross Dingsbums Firma GmbH

Dear Mr. Schultz:

What a stroke of luck it was to run into you at the industry conference in Strasbourg the other day. Who would have guessed that just putting our heads together for five minutes over a cup of coffee would lead to such a brainstorm. I think you could safely say that our backgrounds weren't exactly cut from the same cloth. But lock us in a room together for 15 minutes, and things definitely begin to click.

I was hoping you could lend me a hand with another matter. While the standards we discussed are conventional according to the needs of most markets, the technology in this field is racing ahead like a charging bull. Is there any assurance that if we form a joint venture, connectibilty and, more important, full integration of our products will still be achievable in a reasonable length of time?

Best regards,
John Smith
The Great Widget Group PLC

Hans Schultz
Gross Dingsbums Firma GmbH

Sehr Geehrter Herr Schultz:

Was ein Schlag des Glücks, den es in Sie an der Industrietagung führen sollte, in Sträburg der andere Tag. Wer hägeraten dies nur unsere Köpfe plazieren Zusammen würden Minuten über einer Tasse Kaffee für fünf zu solch einem Geistesblitz führen. Ich denke, dä Sie sicher sagen konnten, dä unsere Hintergründe nicht genau von geschnitten wurden, d Gleiches Tuch. Aber verriegeln Sie uns in einem Zimmer zusammen 15 Minuten, und Dinge beginnen bestimmt zu klicken.

Ich Hoffte, Dä Sie Mir Eine gibt Mit Einer Anderen Angelegenheit Leihen Konnten. While die Standards Wir erörterten sind konventionall entsprechend dem Bedarf der meisten Märkte, d Technologie in diesem Feld macht ein Wettrennen voraus wie eine rennende Bulle. Gibt es irgendwelch? Versicherung das wenn wir formen einen Joint Venture, Connectibility und, wichtiger, Volle Integration unserer Produkte,wird immer noch in einer vernünftigen Länge der Zeit erzielbar sein?

Beste Grü@e,
John Schmied
The gro@e Widget Gruppenag

Hans Schultz
Great what'sit bang company company with limited liability

Dear Mr Schultz:

What a blow of luck, this one it should lead the other day out of you at the industry conference over street castle. Who would have looked after, that, put our heads about a cup of coffee only five minutes, supply a spirit lightning suchly would. I think that you could for certain say that our backgrounds exactly weren't cut by the same cloth. But lock you for us in a room with 15 minutes and things starts clicking certainly.

I hoped that you could borrow a hand with another matter. We discussed the standards lasting conventional according to the need of most markets is, the technology makes in this field a race like one reindeer end bull ahead. This is very insurance this if we form one joint ventures, Connectibilty, and full integration of our products still will more importantly be obtainable in a sensible length of the time?

Best greetings,
John Schmied
Den die gro

Source: Convergence Magazine, Wall Street Journal Europe

messages from a cellphone in both countries is cheap and hugely popular with the impecunious young. But typing a written message on a
cellphone only seven or eight lines deep requires masterly condensa-
tion, and the buttons are hard to hit, even for nimble fingers. So young
Brits insert numbers to save on letters, as in C U 2moro. Vodafone,
Britain's giant mobile-telephone company, offers a selection for the
uninitiated on its Web site.

Young Japanese luckily already use a slang that condenses words and
so is ideal for the telephone. Japanese schoolgirls have a bizarre vocab-
ulary of *"kogaru"* words (*"ko"* meaning "little one" and *"garu"* being
the Japanization of "girl") that involves dropping most of the middle
characters in compounds and then dovetailing the first and last sounds
together to form a whole new word. It is perfect for condensing messages. Young Japanese also use their own "emoticons," which they call
kaomoji. While Western emoticons, long used by e-mailers, read emotions from the mouth, with :-) for a smile or ;-) for a wink, Japanese
concentrate on the eyes. A happy face may be (^ - ^) or (^ o ^), and
embarrassment is (^ ^ ;) with the semi-colon denoting sweat.[31] Thus
do youngsters the world over create their own cultures and languages,
even in a world of inexpensive global communications.

Cultural protection

Some countries, especially where English is not the first language,
dread the erosion of their own cultures in the face of an onslaught of
American products. For years, continental European countries have
subsidized their film and television industries, partly to keep them alive
and partly to fend off the power of Hollywood. In both, they have
failed: indeed, American films take about 80 percent of box office
receipts in much of Western Europe, while the countries of the European Union take 60 percent of American film exports.[32]

However, one effect of the electronic revolution will be to increase
the output of locally made entertainment, for two reasons. First,
demand is likely to rise, as the coming of digital television expands the
number of channels to fill, and television viewers, unlike movie audiences, generally seem to prefer locally made fare. Second, the cost of
supply is declining as digital equipment sharply reduces the cost of

making movies and television programs. True, the big costs of Hollywood movie-making are not technical: they include multimillion dollar payments to stars, directors, and scriptwriters as well as the immense expense of global marketing and distribution. But lower production costs will make it easier to foster local talent.

A likely consequence of these developments will be the re-emergence in Europe of a second-tier industry of national movies and television programs. This industry may not produce many films as good as *The Full Monty*, let alone *Four Weddings and a Funeral*. As with the vast output of Bombay's Bollywood, which produces more movies than any other city on earth, few will be good enough to export to a world market, although some may enjoy regional success. But digital broadcasting will allow people to choose the language of the soundtrack for whatever they watch, and digital manipulation of filmed images will improve the quality of dubbing. Take away the state subsidies, and France and Italy will once again become centers of cinematic excellence, producing the likes of *Les Quatre Cent Coups* and *La Dolce Vita* for discriminating audiences around the world.

Winners and Losers

. . . . .

The new world of electronic communications will include winners and losers, haves and have-nots. In rich countries, governments fret that some groups will be excluded because they are too poor to afford the equipment and gadgets. In poor countries, politicians worry that the nation may not be connected at all. In fact, this might eventually be a revolution of inclusion and opportunity for many of today's have-nots.

In the rich world

For the moment, a "digital divide" exists. A rough rule of thumb, calculated by the OECD, is that, for every $10,000 increase in household income, the number of homes owning a computer rises by seven percentage points.[33]

. . . . .

A series of surveys carried out by the Department of Commerce in the United States since 1995 tells a similar tale. Although access to the Internet has been increasing in all demographic groups, a report released in 1999, called "Falling Through the Net III," found that people with college degrees were eight times more likely to have a PC at home and sixteen times more likely to have Internet access at home than those with an elementary school education. White children were more likely to have access than those in black or Hispanic families; rich city dwellers more likely than poor rural folk; and able-bodied people more likely than disabled. In short, the very people with most to gain were the least likely to have access.[34] However the same survey, a year later, found the fastest growth in Internet access in rural and non-white homes, and among older people.

In time, the divide may shrink further. It is certainly narrower for other electronic technologies, such as games machines. Indeed, the simplicity and usefulness of the mobile telephone, and the variety of ways to pay for it, mean that few complain of a digital divide for this device.

The costs of access to the digital world are falling and convenience is rising. Moreover, the benefits of access will grow. Prices are dropping at a staggering rate for almost all the machinery that connects people. Mobile telephones, set-top boxes for digital television, home computers, videophones: all are becoming cheaper not just relative to other products, but in absolute terms. An Internet-ready PC can now be bought for less than the price of most television sets, which are standard items in almost every home. In addition, the cost of connecting people is tumbling. The only area where the poor may be excluded, for a while, is in the upgrading of telephone and cable networks to carry high-speed computer links. Companies tend to concentrate first on well-to-do areas. But access is still available to the poor, as to many other people, over a standard telephone line. Finally, many of the services now available only with a PC will soon be widely available on Internet-enabled mobile telephones, and on other devices too. Some will be extremely simple to use, such as the Web kiosks with touch screens that are to be found in cities such as Toronto and Singapore.

At the same time, there will be incentives to help those on the wrong side of the digital divide to leap across it. Governments and companies will make huge cost savings by delivering services electronically, so it will be in governments' interest to spend some of those savings to connect the unconnected. When that happens, the beneficiaries will include the housebound, both the old and the disabled. For them, anything that makes communications easier and less expensive is good. The proportion of old people will grow steeply in the rich countries: in several, one person in five will be over sixty-five by 2010. The lives of these people, many of whom will live well into their eighties, will be impoverished by the extent to which modern societies depend on the automobile. Once they become too old to drive, they will find themselves more isolated than elderly people were in the first half of the twentieth century. For these people, many of whose relatives live too far away to visit frequently, inexpensive communications will be a lifeline.

The poor may also find themselves more included in the new electronic society than they were in the old one. A program sponsored by the American government and initially called Making Healthy Music is one of several interesting experiments to link low-income people into community information systems. Participants in low-income homes in Newark, New Jersey, receive a computer, software, and training in exchange for their commitment to try to improve communications with their neighbors. They are then hooked into an intranet linking them to other local users (as well as doctors and health officials). Some proponents of the program claim it has helped participants to build friendships in neighborhoods where people fear to venture into the streets.[35]

In general, though, the uneducated and old benefit less than the well educated and young. Those without education find it hardest to make use of the Internet's many options. Access to all the books in the Library of Congress is of little use if you cannot read. As for older people, they will remain for a generation at a disadvantage to the young, many of whom will have a natural feel for the technology that is essential to the way offices work.

In some countries, this alone is a powerful force for change. In Japanese companies, where age has for so long been tantamount to seniority, older workers suddenly find themselves forced to ask their juniors how to use the most important piece of office equipment, the computer; and junior workers find they no longer need to pass a memo to a senior manager through their immediate boss, with all the opportunities for obstruction that provides, if they can send their comments

· · · · ·

electronically. These new procedures reinforce trends that are already breaking down Japan's hierarchical employment patterns.

The young will grow up with the idea of almost limitless choice of entertainment, of easy access to information, and of the screen and computer as gateways to the rest of the world. The electronic world will be dominated by them.

In the developing world

In rich countries, some members of society may be excluded from the electronic world. But some poor countries may be excluded altogether. Rich countries account for only 15 percent of the world's population, but for 90 percent of global spending on information technology and 80 percent of the world's Internet users. When a computer in a country as poor as Bangladesh costs the equivalent of eight years' pay, how can the divide ever be bridged?[36] [Fig 11-3]

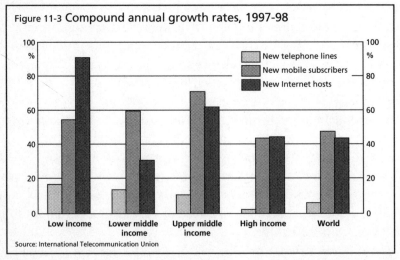

Figure 11-3 Compound annual growth rates, 1997-98

Source: International Telecommunication Union

In fact, as the chart shows, growth has sometimes been faster in poor countries than in rich. To some extent, poor countries control their own destinies. The number of telephone lines per hundred inhabitants has, until now, generally been correlated with a country's level of wealth; wealth seems to be an essential precondition for increasing communications facilities. Yet the effects of poverty are frequently compounded by badly designed telecommunications regulations and by restrictions on would-be private investors, both of which are often designed to shore up an inefficient national monopoly. That increases waiting lists for telephones and drives up the cost of access to the Internet. In Mexico, twenty hours of access a month costs $90, the equivalent of 15 percent of average income; in the United States, the cost is $25, equivalent to 1 percent of income.[37]

Such problems can largely be solved with appropriate policies. Even in the poorest countries, there are wide differences in telephone and Internet use. India, for instance, has thirty telephone main lines, three mobile-telephone subscribers, and four Internet users for every thousand people; in China, the equivalent figures are 110, fifty-four, and sixteen. [38] Poverty alone cannot explain why China should have four times as many Internet users as India does.

Once good policies are in place, developing countries have a great opportunity. They have the potential to skip several stages of technological development and go straight to the most up-to-date networks. Robert Anderson, a white Kenyan who farms 350 acres on the Ugandan border, originally grew apples. The opening of South Africa after the end of apartheid killed that business. Mr Anderson switched to growing roses to sell on the Dutch flower market, employing 540 people. Then he installed a microwave telephone to allow him faster contact with his European buyers, and was able to expand production and employ a further thousand people. The telephone, he told a BBC interviewer, "is what has enabled the whole expansion. Now I can talk to the world."[39]

There will be similar tales in many developing countries. Because developing countries start off with so much less capital investment per worker than rich countries, they have huge scope to grow rapidly simply by buying technology already invented in the rich world and copying first-world production methods. Moreover, while it took decades for previous technologies, such as railways and electricity, to move from the rich world to the poor, communications and computers travel faster. Their border-hopping properties help to bring new ideas to receptive poor countries almost as quickly as they spread through the rich world.

Connected to the world's networks, poor countries can go in a single leap from no communications to the benefits of present-day technology. A village acquires a public telephone: suddenly, its farmers can discover the price for which their goods are being sold from week to week in city markets. A long-distance truck driver acquires a mobile telephone: suddenly, he can double his output by finding a return load. Even without the benefits of the Internet, the telephone can turbocharge development.

Moreover, poor countries have one enormous potential advantage. They are home to most of the world's young. If these countries allow their communications and media industries to flourish – if they liberalize markets, regulate the inevitable concentrations of market power, protect freedom of speech, and promote education and literacy – they will eventually be the biggest beneficiaries of the telecommunications revolution. A country such as India, with enormous creativity and widespread use of English, or Chile, with its relatively open telecommunications market, or China, with its extraordinary attention to education, might leapfrog ahead of many competitors from the rich world. Distance will no longer be an obstacle.

The most important effect of the death of distance will be to narrow gaps, not widen them. Where countries adapt their policies to allow new communications industries to flourish, they will find that the electronic world creates opportunities rather than suppressing them and opens doors rather than shutting them. Communication, after all, is about bridging divides. It's good to talk.

Chapter 1

1. No standard definition exists of the number of Internet users. Like the number of telephone users, it is anybody's guess. At the end of 1999, the OECD calculates that there were 120 million Internet subscribers in the OECD countries. But one firm or family may have several users. In addition, frequency and type of use may vary. Estimates for developing countries are particularly unreliable. A study of America's 199 million adults (at *www.thestandard.com*) found that eighty-six million had used the Internet in the previous thirty days. Of those, forty-six million had used it from home only; twenty-one million from work only; and the rest from more than one location.

   The *Computer Industry Almanac* estimates that there were 259 million users worldwide at the end of 1999 and forecast 374 million at the end of 2000. NUA, an Irish consultancy, produces a global estimate from an amalgamation of other estimates. At the end of June 2000, this produced a total of 330 million. This book sticks, for the sake of simplicity, with an estimate by the International Telecommunication Union (ITU) of 385 million users in 2000.

2. The title of a survey of telecommunications by Adrian Wooldridge, *The Economist*, 9 October 1999.

3. James Barrett of Palmer & Dodge, speaking at "Globalization of E-Commerce," Harvard.Net.news, 2 June 2000; "The Next Step: Charting E-Commerce's Future," *International Herald Tribune*, 19 June 2000.

4. Peter Bradshaw and Henry Blodget, *Internet/E-Commerce Quarterly Handbook*, Q1 2000, Merrill Lynch.

5. Alfred Marshall, *Principles of Economics*, 8th ed. (1920, book IV).

6. "Non-English Domain Names to Come On-line," *Wall Street Journal Europe*, 10–11 November 2000. As a result, the original thirty-seven characters used for Web addresses – the twenty-six letters of the English alphabet, ten numerals, and a hyphen – were joined by more than forty billion Asian character sets. Users still have to type ".com" in Roman letters.

7. Quoted in Pam Woodall, "Untangling the E-conomy," *The Economist*, September 23 2000.

· · · · ·

8. Laszlo Solymar, *Getting the Message* (Oxford: Oxford University Press, 1999), 51. To be fair to the Admiralty, it had just come to the end of a war with France and was thus less interested than it might have been in such innovations. Besides, it had already invested in developing an alternative technology, the optical telegraph.

9. Elliot Wilson, "Sweden's Winning Formula," *Net Profit*, May 2000, 10.

10. Working Party on Telecommunication and Information Services Policies, "Local Access Pricing and E-Commerce," (Paris: OECD, July 2000), 9.

11. Catherine L. Mann, Sue E. Eckert, and Sarah Cleeland Knight, *Global Electronic Commerce* (Washington, DC: Institute for International Economics, 2000), 22–3.

12. "The Default Language," *The Economist*, 15 May 1999.

13. Andrew Odlyzko, "The History of Communications and its Implications for the Internet." Unpublished preliminary manuscript at *www.research.att.com/~amo/doc.networks.html*.

14. In 2000 Microsoft was developing software to filter incoming electronic information by trading off its importance against a user's likely receptivity. The aim would be to judge the information's relevance against the time of day, whether the user was on the telephone or talking to somebody and so on. But there are snags. Microsoft's earlier attempt to do something similar in 1995, with an on-screen valet called Bob, was widely ridiculed. And Blue Mountain Arts, an Internet greetings card company, sued Microsoft in 1998 because the software company had launched an e-mail filter that directed Blue Mountain's electronic cards straight into the trash can. John Markoff, "Microsoft Goal: 'Traffic Cops' to Fight Data Jams," *International Herald Tribune*, 18 July 2000.

15. *Telecommunications Indicators Update April–May–June 2000* (Geneva: International Telecommunication Union, 2000).

Chapter 2

1. Douglass North, "Ocean Freight Rates and Economic Development 1750–1913," *Journal of Economic History*, 18, no. 4, 537–55.

2. Richard Mullen, *Anthony Trollope: A Victorian in his World* (London: Duckworth, 1990), 85–6.

3. This chapter (and some others) have been strongly influenced by Andrew Odlyzko, "The History of Communications and its Implications for the Internet", which can be found at *www.research.att.com/~amo/doc.networks.html*.

4. Victoria Glendinning, *Trollope* (London: Pimlico, 1992), 197. They also struck a blow for novelists such as Trollope himself. So superb was the Victorian Post Office that, by the 1860s, readers in the London area could despatch a letter ordering the latest novel from Charles Mudie's famous circulating library and receive their copy the same day. Richard Mullen, *Anthony Trollope: A Victorian in his World* (London: Duckworth, 1990), 162.

5. Tom Standage, *The Victorian Internet* (London: Phoenix, 1999) is the best history of the telegraph and of the many similarities between it and the Internet.

6. The phrase was coined by Clayton Christensen in *The Innovator's Dilemma* (Cambridge, Mass.: Harvard Business School Press, 1997).

7. The capacity of such devices, calculates Laszlo Solymar, was 0.2 bits per second, against the ten billion carried today by a single optical fiber. *Getting the Message: A History of Communications* (Oxford: Oxford University Press, 1999), 29.

8. Odlyzko, "The History of Communications and its Implications for the Internet."

9. Solymar, *Getting the Message: A History of Communications*, 3.

10. Odlyzko, "The History of Communications and its Implications for the Internet."

11. Quoted in Standage, *The Victorian Internet*, 156.

12. Alfred Marshall, *The Principles of Economics*, 8th ed. (1920, book IV).

13. Geoffrey Blainey, *The Tyranny of Distance* (Sydney: Pan Macmillan, 1966), 225.

14. Solymar, *Getting the Message: A History of Communications*, 82–3.

15. Blainey, *The Tyranny of Distance*, 223.

16. Stephen Goodwin, "Trapped On Everest? I'm On My Mobile, "The *Independent*, 27 April 1996. Eamon Fullen's rescue began, as newspapers reported at the time, with "the standard procedure for males in distress – call the wife." His wife took the call, made from his solar-powered satellite telephone, two thousand miles away in Hong Kong and alerted the Nepalese army, which organized a helicopter rescue. Mobile telephones have become the bane of Scotland's mountain rescue teams because they encourage hill-walkers to ignore the promptings of common sense.

17. Andrew Odlyzko recalls an editor of a learned journal who, up until the mid-1980s, found that the best way to attract the attention of a tardy academic referee was to send him a telegram. Personal communication, August 2000.

18. Pam Woodall, "Untangling the E-conomy," *The Economist*, 23 September 2000.

19. Quoted in Odlyzko, "The History of Communications and its Implications for the Internet."

20. Solymar, *Getting the Message: A History of Communications*, 126–9, reproduces a splendid Art Nouveau advertisement for an opera on Telefon Hirmondo.

21. MMXI/SIFO, quoted in "Sweden's Winning Formula," *Net Profit*, no. 41, May 2000.

22. International Telecommunication Union.

23. Walter B. Wriston, *The Twilight of Sovereignty: How the Information Revolution is Transforming Our World* (New York: Scribner's, 1992), 36.

24. Robert J. Gordon, "Does the 'New Economy' Measure up to the Great Inventions of the Past?" *NBER Working Paper No. W7833, August 2000*.

25. Robert H. Frank and Philip J. Cook, *The Winner-Take-All Society: How More and More Americans Compete for Ever Fewer and Bigger Prizes, Encouraging Economic Waste, Income Inequality, and an Impoverished Cultural Life* (New York: The Free Press, 1995).

26. *Defining Moments* (London: A.T. Kearney, private publication, 1996), 38.

27. Quoted by Diane Coyle, Merrill Lynch/Imperial College Lecture, 10 November 1999. Available at *http://dialspace.dial.pipex.com/diane.coyle/credo.htm*.

28. "War Games," *The Economist*, 22 April 2000. The Japanese authorities had been embarrassed in 1998 when the radar and global-positioning equipment found in a captured North Korean submarine turned out to be based on popular gadgets made by Japanese consumer-electronics firms.

29. Standing for the Advanced Projects Research Agency of the US Defense Department.

30. Quoted in John Naughton, *A Brief History of the Future: The Origins of the Internet* (London: Weidenfeld & Nicolson, 1999), 232.

31. Quoted in *Ibid.*, 248.

32. Although, of course, the world's population has multiplied threefold since the coming of television.

33. *Challenges to the Network: Internet for Development* (Geneva: International Telecommunication Union, 1999).

34. *Ibid.* Traffic measured by the number of Web sites visited by each host computer.

35. Amazon accounts for about eight of every ten books sold on-line in the United States.

36. M. Karshenas and Paul Stoneman "Rank, stock order and epidemic effects in the diffusion of new process technology," *Rand Journal of Economics* 24, 4, 503–528, Winter 1993.

37. Pam Woodall, *"Untangling the E-conomy,"* *The Economist*, 23 September 2000.

38. Paul A. David, "The Dynamo and the Computer: A Historical Perspective on the Modern Productivity Paradox," *The American Economic Review*, May 1990.

39. See Coyle, Merrill Lynch/Imperial College Lecture.

Chapter 3

1. *World Telecommunication Development Report 1999* (Geneva: International Telecommunication Union, 1999).

2. *International Bandwidth 2000* (Washington, DC: TeleGeography, 2000).

3. *Ibid.*

4. *Ibid.*

5. *Direction of Traffic* (Geneva and Washington, DC: International Telecommunication Union and TeleGeography), 62.

6. *Trends in Telecommunication Reform* (Geneva: International Telecommunication Union, 1999).

7. *Direction of Traffic*, 51.

8. Michael Minges, personal communication, 24 August 2000.

9. *World Telecommunication Development Report 1999*.

10. "When India Wires Up," *The Economist*, 22 July 2000.

11. Adrian Wooldridge, "The World in Your Pocket," *The Economist*, 9 October 2000.

12. Working Party on Telecommunication and Information Services Policies, *Cellular Mobile Pricing Structures and Trends* (Paris: OECD, May 2000),19–29.

13. Wooldridge, "The World in Your Pocket."

14. *Cellular Mobile Pricing Structures and Trends*, 29.

15. Survey in June 2000. See *www.oftel.gov.uk*. Only 9 percent of those over seventy-five had a mobile telephone, compared with 66 percent of fifteen to thirty-four-year-olds.

16. *Ibid.*, 11.

17. Wooldridge, "The World in Your Pocket."

18. *Ibid.*

19. *Ibid.*

20. "Morocco Goes Mobile," *The Economist*, 13 May 2000.

21. *World Telecommunication Development Report 1999*, 69.

22. Capacity on America's wired network is designed to carry calls on the two peak days of traffic: Mother's Day, and the Monday after Thanksgiving weekend.

23. *Telecommunication Indicators Update, April–May–June 2000* (Geneva: International Telecommunication Union, 2000).

24. *Cellular Mobile Pricing Structures and Trends*, 63.

25. Bill Bryson, *Made in America: An Informal History of the English Language in the United States* (New York: William Morrow, 1995).

26. In the final years of President Mobutu, before Zaire became the Congo and dissolved into chaos, its broadcasting authority used to begin its evening news program showing the president descending to earth from the clouds.

27. The average capacity of European cable systems is twenty-seven channels; cited in "Europe's 'Other' Channels," *Screen Digest*, March 1997, 57.

28. Richard Kee, John Davison, Mari Vahanissi, and Kate Hewett, *Cable: The Emerging Force in Telecoms and Interactive Markets* (London: Ovum, 1996).

29. *World Telecommunication Development Report 1998* (Geneva: International Telecommunication Union, 1998).

30. For instance, British children aged two to nine in homes with cable or satellite devoted 64 percent of their viewing to cable or satellite channels in October 1995; the share for adults in such homes was 36 percent. John Clemens and Jane Key, *Trends in Viewing in Cable TV Homes 1990 – 95* (London: Independent Television Commission, 1996), 7.

31. "The Fright After Christmas," *The Economist*, 5 February 2000.

32. Veronis, Suhler and Associates, *Communications Industry Forecast*, 53.

33. "Murdoch's Empire," *The Economist*, 9 March 1996, 101.

34. Emma Duncan, "Wheel of Fortune: A Survey of Technology and Entertainment," *The Economist*, 21 November 1998.

35. Christopher Dunkley, "Broadcasting's Bruiser," *Financial Times*, 26–27 August 2000.

36. *World Telecommunication Development Report 1998*.

37. Anthony DePalma, "Wall St. Uses Webcasts to Give Opinions Directly to Investors," *New York Times*, 27 March 2000.

38. Working Party on Telecommunication and Information Services Policies, *Local Access Pricing and E-Commerce* (Paris: OECD, July 2000), 21.

39. Quoted in Duncan, "Wheel of Fortune: A Survey of Technology and Entertainment."

Chapter 4

1. See note 1, Introduction.

2. John Naughton, *A Brief History of the Future: The Origins of the Internet* (London: Weidenfeld & Nicolson, 1999) xii.

3. *Ibid.*, 148.

4. Katie Hafner and Matthew Lyon, *Where Wizards Stay Up Late* (New York: Simon & Schuster, 1996).

5. A point proved by my daughter when she visited the remote (but Internet-connected) island of Lamu in Kenya in 1998.

6. *www.CyberAtlas.internet.com*.

7. Alaina Kanfer, "It's a Thin World: The Association between E-mail Use and Patterns of Communication and Relationships," 10 March 1999. Report available at *www.ncsa.uiuc.edu/edu/trg/email*

8. Vanessa Houlder, "Failing to Get the Message," *Financial Times*, 17 March 1997.

9. "Trying to Connect You," *The Economist*, 24 June 2000.

10. Amy Cortese, "A Way out of the Web Maze," *Business Week*, 27 February 1997.

11. See *http://nuevaschool.org/* for one "advice engine."

12. Although it might also have done damage, if some of the technicians purporting to cure the bug had left nasty surprises in the software.

13. "The Consensus Machine," *The Economist*, 10 June 2000.

14. "Optic Verve," *The Economist*, 15 July 2000.

15. Working Party on Telecommunication and Information Services Policies, *Local Access Pricing and E-Commerce*, (Paris: OECD, July 2000), 23 and 28.

16. "After the PC," *The Economist* 12 September 1998.

17. Neil Gershenfeld, *When Things Start to Think*, (New York: Henry Holt and Company, 1999).

18. Julie V. Iovine, "Can a House be a Little Too Smart?" *International Herald Tribune*, 17 January 2000.

19. *Net Profit*, no. 39, March 2000.

20. Carol J. Williams, "Cooking and Cleaning at the Touch of a Button," *International Herald Tribune*, 17 January 2000.

21. Surveys by Arbitron and Edison Media Research. Quoted in Working Party on Telecommunication and Information Services Policies, *Local Access Pricing and E-Commerce* (Paris: OECD, July 2000), 24.

22. Andrew Odlyzko, *The History of Communications and its Implications for the Internet*, 73.

23. *Ibid.*, 71.

24. *Ibid.*, 71–72.

25. Victoria Shannon, "Oracle Chief Foresees European Internet Lead," *International Herald Tribune*, 13 October 1999.

26. Quoted by Thomas L. Friedman in "Internet Entrepreneur: Act Small, Think Global," *International Herald Tribune*, 10 March 1999.

27. Quoted in "The Death of Distance: A Survey of Telecommunications," *The Economist*, 30 September 1995.

Chapter 5

1. John Peet, "Shopping Around the Web", *The Economist*, 26 February 2000.

2. Matthew R. Sanders, "Global eCommerce Approaches Hypergrowth," Forrester Research, 18 April 2000.

3. This view was convincingly put by Tim Jackson in "When Choice Beats Price," *Financial Times*, 17 October 2000. "Helping customers find the lowest price is helpful," he argued, "but helping them find something they otherwise could not find at all is terrific."

4. Peet, "Shopping Around the Web."

5. "Europe's dot.bombs," *The Economist*, 5 August 2000.

6. Forrester Research forecasts global on-line porn sales at $336 million in 2001. Others think the true figure may be three times that.

7. "The State of On-line Retailing," Shop.org and The Boston Consulting Group, 2000.

8. Philip Evans and Thomas S. Wurster, *Blown to Bits: How The New Economics of Information Transforms Strategy* (Boston: Harvard Business School Press, 1999).

9. Peet, "Shopping Around the Web."

10. Evans and Wurster, *Blown to Bits: How the New Economics of Information Transforms Strategy*, 1–4.

11. "The State of On-line Retailing."

12. "All Yours," *The Economist*, 1 April 2000, and author's conversations with Richard Gerstein, Reflect.

13. Catherine L. Mann, Sue E. Eckert, and Sarah Cleeland Knight, *Global Electronic Commerce* (Washington, DC: Institute for International Economics, 2000), 62.

14. Aline Sullivan, "Tickets On-line? Buyers Hesitate," *International Herald Tribune*, 21 July 2000.

15. Mann *et al.*, *Global Electronic Commerce*, 62.

16. "E-cash 2.0," *The Economist*, 19 February 2000.

17. *The Economic and Social Impacts of Electronic Commerce* (Paris: OECD, 1999), 63

18. *Ibid.*, 63–64.

19. "On-line, Off-course," *The Economist*, 10 June 2000.

20. Jack Schofield, "Drat, It's Down Again," *Guardian*, 27 July 2000.

21. Peet, "Shopping Around the Web."

22. "Driving Down Delivery Times from Days to Minutes," *Net Profit*, number 42 June 2000.

23. "Dotty About Dot.commerce?" *The Economist*, 10 June 2000.

24. Peter Bradshaw and Henry Blodget, *Internet/ E-Commerce Quarterly Handbook*, Q3 2000, Merrill Lynch.

25. Adrian Wooldridge, "The World in Your Pocket," *The Economist*, 9 October 1999.

26. Forrester Research, *The E-mail Marketing Dialogue*, January 2000.

27. Saul Hansell, "Ineffective and Costly, Targeted Ads on the Internet Miss the Mark," *International Herald Tribune*, 8 May 2000.

28. "Tobacco Firms Walk the On-line Tightrope," *Net Profit*, no. 41, May 2000.

29. Michael Maubossin, "Absolute Power: The Internet's Hidden Order," Credit Suisse First Boston, 20 December 1999.

30. "The State of On-line Retailing."

31. Erik Brynjolfsson and Michael D. Smith, "Frictionless Commerce? A Comparison of Internet and Conventional Retailers," *Management Science*, vol. 46, no.4, April 2000.

32. Michael D. Smith and Erik Brynjolfsson "Understanding Digital Markets: Review and Assessment," September 1999. Available from *http://ecommerce. mit.edu/papers/ude* working paper.

33. "Web Wonders: the economic impact of the Internet", HSBC, January 2000.

34. See "Pricing Gets Personal," The Forrester Report, April 2000.

35. Peet, "Shopping Around the Web."

36. "European Buyers Flatten Prices," *Net Profit*, no. 35, November 1999.

37. Gaston F. Ceron, "A Breakthrough in Web Investing," *Wall Street Journal*, 14 August 2000.

38. "Tobacco Seller Flourishes On-line" and "Electronic Boost for Rural Craftsmen," *Net Profit*, no. 40, April 2000.

39. "Designs On Your Bike and Fitting Ways with Shoes," *Net Profit*, no. 35, November 1999.

40. B. Joseph Pine II, *Mass Customization: The New Frontier in Business Competition* (Boston: Harvard Business School Press, 1999).

41. *Challenges to the Network: Internet for Development* (Geneva: International Telecommunication Union, 1999), 45, figure 32.

42. *Local Access and E-Commerce Pricing* (Paris: OECD, 2000), 16.

43. "The State of On-line Retailing."

44. *Net Profit*, no. 40, April 2000.

45. Richard S. Tedlow, "Roadkill on the Information Superhighway," *Harvard Business Review*, November/December 1996, 15.

46. "Net Route to Bank Expansions," *Net Profit*, no. 39, March 2000.

Chapter 6

1. Quoted in Martin Brookes and Zaki Wahhaj, "*Is the Internet Better than Electricity?* (London: Goldman Sachs, 2 August 2000).

2. *The Economic and Social Impact of Electronic Commerce* (Paris: OECD, 1999), 11.

3. *A New Economy? The Changing Role of Innovation and Information Technology in Growth* (Paris: OECD, 2000), 50.

4. *The Economic and Social Impact of Electronic Commerce*, 14.

5. *Ibid.*, 61.

6. *Ibid.*, 61.

7. "A Spoonful of Sugar," *The Economist*, 1 July 2000.

8. "Web Wonders: The Economic Impact of the Internet," HSBC, January 2000.

9. Nick Valery, bureau chief of *The Economist* in Tokyo, personal communication.

10. David Bowen, "Seamless Supply Chains," *Net Profit*, no. 38, February 2000.

11. "New Wiring," *The Economist*, 15 January 2000.

12. Ray Hurst, "Dispersed Experts Collaborate On-line," *Net Profit*, no. 39, March 2000.

13. Ibid.

14. Ben Schiller, "Tapping Remote Knowledge," *Net Profit*, no. 38, February 2000.

15. *A New Economy? The Changing Role of Innovation and Information Technology in Growth*, 47.

16. *Ibid.*, 54.

17. *Ibid.*, 44.

18. "Building a New Boeing," *The Economist*, 12 August 2000.

19. *A New Economy? The Changing Role of Innovation and Information Technology in Growth*, 83.

20. Kevin Done, "Airlines Take On-line Route to Soft Landing," *Financial Times*, 24 February 2000.

21. Carlos Grande, "US Groups Form Net Buyers' Pool," *Financial Times*, 11 January 2000.

22. Emma Charlton, *European Electronic Procurement* (London: Net Profit Publications, 2000), 33.

23. Charlton, *European Electronic Procurement*, 11.

24. Kent Brittan, United Technologies Corp, conversation with author, June 2000.

25. Sam Kinney, *An Overview of B2B Purchasing Technology: Response to Call for Submissions* (Washington, DC: Federal Trade Commission, 2000)

26. Ian Fletcher, UK Department of Trade and Industry, personal communication with the author, August 2000.

27. Matthew Symonds, "The Net Imperative: A Survey of Business and the Internet," *The Economist*, 26 June 1999.

28. Charlton, *European Electronic Procurement*, 37.

29. Symonds, "The Net Imperative: A Survey of Business and the Internet."

30. Charlton, *European Electronic Procurement*, 40–41.

31. Armand V. Feigenbaum, "Net Rewrites Rules of Production," *Financial Times*, 9 May 2000.

32. "All Yours," *The Economist*, 1 April 2000.

33. Frances Cairncross, "Inside the Machine", *The Economist*, 11 November 2000.

34. "All Yours," *The Economist*, 1 April 2000.

35. *The Economic and Social Impact of Electronic Commerce* (Paris: OECD, 1999), 62.

36. *A New Economy? The Changing Role of Innovation and Information Technology in Growth*, 55–56.

· · · · ·

37. On some estimates, 80 percent of visitors to corporate Web sites go to the careers page. Nuala Moran, "Employers Switch from Print to Internet Adverts," *Financial Times*, 3 May 2000.

38. Alan B. Krueger, "Net Job Searches Change the Market and the Economy," *International Herald Tribune*, 24 July 2000.

39. *A New Economy? The Changing Role of Innovation and Information Technology in Growth*, 46.

40. *Ibid.*, 47.

41. Charlton, *European Electronic Procurement*, 42.

42. Frances Cairncross, "The Best ... and the Rest: A Survey of Global Pay," *The Economist*, 8 May 1999.

43. Ben Schiller, "Web Co-operation and Communication Smoothes Product Development Processes," *Net Profit*, no. 35, November 1999.

44. *A New Economy? The Changing Role of Innovation and Information Technology in Growth, Paris*, 36.

45. Schiller, "Web Co-operation and Communication Smoothes Product Development Processes."

46. Ben Schiller, "Smoothing Customer Service," *Net Profit*, no. 39, March 2000.

47. A phrase coined by Ward Saloner of Stanford Graduate Business School. Frances Cairncross, "Inside the Machine," *The Economist*, 11 November 2000

48. *A New Economy? The Changing Role of Innovation and Information Technology in Growth*, 9.

49. *Ibid.*, 38–39.

Chapter 7

1. See *www.ecitizen.gov.sg*, mentioned by Matthew Symonds, "The Next Revolution: A Survey of Government and the Internet," *The Economist*, 24 June 2000.

2. His words paraphrased by Neal Pierce, "Now an E-Government Revolution," *International Herald Tribune*, 9 August 2000.

3. Eli M. Noam, "Electronics and the Dim Future of the University," *Science*, 13 October 1995, 247–9.

4. Doug Struck, "Blacklisted on the Web, 58 Candidates Lose in South Korean Vote," *International Herald Tribune*, 15–16 April 2000.

5. Sverker Lindbo, e-mail to author, April 1997.

6. James S. Fishkin, *Democracy and Deliberation* (New Haven, Conn: Yale University Press, 1991), 67.

7. Symonds, "The Next Revolution: A Survey of Government and the Internet."

8. Quoted in *Ibid*.

9. Alvin and Heidi Toffler, *Creating a New Civilization: The Politics of the Third Wave* (Atlanta: Turner Publishing, 1995).

10. David Butler and Austin Ranney, eds., *Referendums Around the World: The Growing Use of Direct Democracy* (New York: Macmillan, 1994), 5.

11. Jeremy Sharrard, "Sizing US E-Government," The Forrester Report, August 2000.
12. *Ibid.*
13. *Ibid.*
14. Michael Nelson, FCC, personal communication with author, 2 May 1997.
15. See *www.servicearizona.ihost.com*.
16. *www.sii.cl*.
17. Symonds, "The Next Revolution: A Survey of Government and the Internet." See *www.naestvednet.dk*.
18. *www.maxi.com.au*.
19. *Washington Post*, 5 April 2000.
20. "Bar-Coding the Poor," *The Economist*, 25 January 1997.
21. "Digital-al Cash," *The Economist*, 15 June 1996.
22. David Brin, "The Transparent Society," *Wired*, December 1996, 62.
23. But such on-line requests can cause havoc. In July 1998, Krystava Schmidt's mother posted an on-line plea for help to find her, after she went missing from her home in Mounds View, a suburb of Minneapolis. Police officers found her soon afterwards – but in 2000 about thirty e-mails and a hundred telephone calls a week offering information on the child were still pouring into the Mounds View police station. Barnaby Fletcher, "The Case is Closed but Not Always on the Internet," *International Herald Tribune*, 17 April 2000.
24. *www.crimenet.com.au*.
25. David Cohen, "Criminals Caught in the Net", *Daily Telegraph*, 1 June 2000.
26. Personal communication, Dr Robert Cox, Hays Medical Center, 5 July 2000.
27. Milt Freudenheim, "Video Approach to Hospital Care," *International Herald Tribune*, 27 February 1997.
28. "Clean Bill of Health for NHS Hospitals," *Financial Times*, 21 March 2000.
29. Alexandra Wyke, "The Future of Medicine," in *Going Digital: How New Technology is Changing Our Lives* (London: The Economist in association with Profile Books, 1996), 250.
30. John Seely Brown and Paul Duguid, *The Social Life of Information* (Boston: Harvard Business School Press), 225.
31. Sir Graeme Davies, Principal of Glasgow University, personal communication to author, 19 May 2000.
32. Noam, "Electronics and the Dim Future," 247–9.
33. "The Doctor Will See You Now – Just Not in Person," *Business Week*, 3 October 1994, 117.
34. Catherine L. Mann, Sue E. Eckert, and Sarah Cleeland Knight, *Global Electronic Commerce* (Washington, DC: Institute for International Economics, July 2000), 80–81.
35. Austan Goolsbee, "In a World Without Borders: The Impact of Taxes on Internet Commerce," *Quarterly Journal of Economics*, vol. 115, no. 2 (May 2000), 561–76.
36. Tim Jackson, "In Praise of Amazon," *Financial Times*, 1 August 2000.

· · · · ·

37. Matthew Bishop, "The Mystery of the Vanishing Taxpayer: A Survey of Global-isation and Tax," *The Economist*, 29 January 2000.

38. Mann *et al. Global Electronic Commerce*, 89.

39. Robert E. Allen, "The Borderless Superpower: Information Technology's Emerging Role in World Politics, Business, and Economic Growth" (speech before the US Council on Foreign Relations, 26 October 1994).

Chapter 8

1. Peter Bradshaw and Henry Blodget, *Internet/ E-Commerce Quarterly Handbook*, Q3 2000, Merrill Lynch.

2. *New York Times* book review, 12 July 1987.

3. "The Internet: A Thinkers' Guide," *The Economist*, 1 April 2000.

4. Gavyn Davies, Martin Brookes, Neil Williams, "Technology, the Internet, and the New Global Economy," Global Economics Paper no. 39, Goldman Sachs, 17 March 2000.

5. Martin Brookes and Zaki Wahhaj, "The Shocking Economic Effects of B2B," Global Economics Paper no. 37, Goldman Sachs, 3 February 2000.

6. Paul David, "The Dynamo and the Computer: An Historical Perspective on the Modern Productivity Paradox," *American Economic Review*, May 1990.

7. Pam Woodall, "Untangling the E-conomy," *The Economist*, 23 September 2000.

8. *Ibid.*

9. Alan Greenspan, remarks before the National Governors' Association, 5 February 1996, quoted in Dr Sushil Wadhwani, Bank of England, "The Impact of the Internet on UK Inflation," speech delivered at the London School of Economics, 23 February, 2000.

10. *The Economic and Social Impact of Electronic Commerce* (Paris: OECD, 1999), 23.

11. Robert J. Gordon, "Has the 'New Economy' Rendered the Productivity Slowdown Obsolete?' 14 June 1999. Available at *http://faculty-web.at.nwu.edu/economics/gordon/334.html*

12. Stephen Oliner and Daniel Sichel "The Resurgence of Growth in the Late 1990s: Is Information Technology the Story?", February 2000.

13. Erik Brynjolfsson and Lorin M. Hitt, "Beyond the Productivity Paradox: Computers are the Catalyst for Bigger Changes," Communications of The ACM, vol 41, no 8, August 1998 49–55.

14. *A New Economy? The Changing Role of Innovation and Information Technology in Growth* (Paris: OECD, 2000).

15. Davies *et al.*, "Technology, the Internet and the New Global Economy."

16. Sushil Wadhwani, op cit.

17. Barclays Capital, "The Great E-scape," 14 December, 1999.

18. Davies *et al.*, "Technology, the Internet, and the New Global Economy."

19. Sarah Chang, "1999 IPO Year-in-Review," *http://ipomaven.123jump.com*.

20. "When the Bubble Bursts," *The Economist*, 30 January 1999.

21. James Lardner: "Ask Radio Historians about the Internet," *www.usnews.com*.

22. Tim Jackson, "Think Again, Oh Sage of Omaha," *Financial Times*, 2 May 2000.

23. World Development Indicators, 2000. The World Bank, Washington, D.C.

24. See, for instance, Danny Quah, "A Weightless Economy," *The UNESCO Courier*, December 1998.

25. See, for instance, Peter Scott of Oracle, quoted in Shailagh Murray, "Rise of Tele-Business in Ireland Gives Jobless in EU New Prospects," *Wall Street Journal Europe*, 5 February 1997.

26. Louis Uchitelle, "Corporate Migration Soars in the US," *International Herald Tribune*, 26 July 2000.

27. "Bangalore Bytes," *The Economist*, 23 March 1996.

28. Peter Marsh, "Using the Net to Get Closer to the Customer," *Financial Times*, 17 February 2000.

29. Ray Hurst, "Dispersed Experts Collaborate On-line," *Net Profit*, no. 39, March 2000.

30. Led by Paul Krugman, *Development, Geography, and Economic Theory, Boston*, (MIT Press, 1995).

31. Pam Woodall, "A Hitchhiker's Guide to Cybernomics: A Survey of the World Economy," *The Economist*, 28 September 1996, 43.

32. Alexandra Wyke, "The Future of Medicine," in *Going Digital: How New Technology is Changing Our Lives* (London: *The Economist* in association with Profile Books, 1996), 242.

33. "All Yours," *The Economist*, 1 April 2000.

34. John Heilemann, "It's the New Economy, Stupid," *Wired*, March 1996, 70.

35. Robert H. Frank and Philip J. Cook, *The Winner-Take-All Society: How More and More Americans Compete for Ever Fewer and Bigger Prizes, Encouraging Economic Waste, Income Inequality, and an Impoverished Cultural Life* (New York: The Free Press, 1995). The idea was first described by Sherwin Rosen in "The Economics of Superstars," *American Economic Review*, 71 (December 1981), 845–58.

36. *A New Economy? The Changing Role of Innovation and Information Technology in Growth* (Paris: OECD, 2000), 28.

37. *Ibid*.

Chapter 9

1. Ithiel de Sola Pool, *Technologies of Freedom* (Cambridge, Mass: Belknap Press, 1983).

2. Michael Meyer, "Whose Internet Is It?" *Newsweek*, 22 April 1996.

3. David Pilling, "Prozac On-line Lifts Drug Sector Spirits," *Financial Times*, 3–4 June 2000.

4. Eager to prove this point, several British journalists have reported on their success in acquiring Viagra – and their embarrassment in subsequently disposing of it. See, for instance, Patti Waldemeir, "E-Practitioners Fell for My On-line Deception," *Financial Times*, 22 June 2000.

5. Catherine L. Mann, Sue E. Eckert, and Sarah Cleeland Knight, *Global Electronic*

Commerce (Washington, DC: Institute for International Economics, July 2000), 121.

6. Gregory C. Staple, ed., "Settlements for Phone Sex," *TeleGeography 1996–97: Global Telecommunications Traffic Statistics and Commentary* (Washington, DC: TeleGeography, 1996).

7. "Heavy Breathing," *The Economist*, 30 July 1994.

8. James Mackintosh, "Internet Access Provider Boosts Efforts to Censor Pornography," *Financial Times*, 6 May 1996.

9. "US Court Overturns Law To Curb Internet," *International Herald Tribune*, 13 June 1996.

10. Mann et al., *Global Electronic Commerce*, 139.

11. Silvia Ascarelli and Kimberley A. Strassel, "German Cases Illuminate Struggle to Regulate Net," *Wall Street Journal Europe*, 21 April 1997.

12. Radio Free Europe/Radio Liberty lists on its Web site "The Twenty Enemies of the Internet:" the worst offenders among forty-five countries that restrict, wholly or partly, their citizens' access to the Internet. In Myanmar, the punishment for owning an undisclosed computer is up to fifteen years in prison.

13. "The Top Shelf," *The Economist*, 18 May 1996.

14. Katharine A. Schmidt, "Babysitter or Big Brother?" *Wall Street Journal Interactive*, 13 June 2000.

15. Although the children may learn to hack their way through. In March 2000 Microsystems Software, which distributes a filter technology called Cyber Patrol that is widely used in American libraries and elementary schools, sued two men for distributing a method for children to discover their parents' password and see a list of the 100,000 or so Internet sites that it screened out. "Filter Busters Sued," *International Herald Tribune*, 20 March 2000.

16. Chris Roth, "The Rise and Fall of the Circulating Library," *www.wiu.edu*.

17. Caroline E. Mayer, "$40 and You're Someone Else," *International Herald Tribune*, 22 May 2000.

18. "Consumers@Shopping: An International Study of Electronic Commerce," Consumers International, September 1999. The test used researchers in Australia, Belgium, Germany, Greece, Hong Kong, Japan, Norway, Spain, Sweden, the United Kingdom, and the United States. It was carried out in late 1998 and early 1999.

19. Brandon Mitchener, "Web Disputes Go On-line," *Wall Street Journal Europe*, 21 March 2000.

20. "The Surveillance Society," *The Economist*, 1 May 1999.

21. *Ibid*.

22. H. Stuart Taylor, letter, "If the Dates Fit, Then Capitalise on Them," *Financial Times*, 1 August 1996.

23. Michael Moss, "Internet Callers Often Know Who's Trying to Click In," *New York Times*, 14 October 1999.

24. Simon Hughes and Paul Thompson, "Mobile Phones Left Trail for Cops," *The Sun*, 5 March 1997.

25. Phil Reeves, "Rumours Run Wild about Dead Leader," *Independent*, 30 April 1996.

26. On acquiring such a pass from the INS office in Newark Airport, I asked anxiously what I should do if I lost it. "It doesn't matter," said the official behind the desk, "just as long as you don't lose your right hand."

27. Lorrie Faith Cranor *et al.*, "Beyond Concern: Understanding Net Users' Attitudes about On-line Privacy." AT&T Research 1999.

28. "On-line Prying Made Easy," *Business Week*, 30 September 1996.

29. "Could Try Harder," *The Economist*, 27 May 2000.

30. "The Surveillance Society," *The Economist*, 1 May 1999.

31. Network Solutions (*www.networksolutions.com*).

32. See *www.icann.org*.

33. "You Name It," letter to *The Economist*, 13 July 1996.

34. "Can You?" *The Economist*, 4 March 2000.

35. *Ibid.*

36. Steven V. Brull, "The Record Industry Can't Stop the Music," *Business Week*, 15 May 2000, 45.

37. "Here, There and Everywhere," *The Economist*, 24 June 2000.

38. de Sola Pool, *Technologies of Freedom*, 249.

39. Contact me at *fac@economist.com*.

40. A large survey in March 2000 by Yankelovich Partners found that two-thirds of consumers said that listening to a song on-line prompted them to buy a CD or cassette featuring the song. Anna Wilde Mathews, "Samplers of Music on Internet Buy CDs in Stores, Study Says," *Wall Street Journal Interactive*, 15 June 2000.

41. David Segal, "Music on the Web: A 'Scourge' that Could Bring More Riches than Ever," *International Herald Tribune*, 23 June 2000.

42. Neil Strauss, "The Downloading of Music Goes Mainstream," *International Herald Tribune*, 7 April 1999.

Chapter 10

1. Richard Tomkins, "Coca-Cola Loses its Fizz," *Financial Times*, 18 July 2000.

2. John Peet, "Shopping Around the Web," *The Economist*, 26 February 2000.

3. *www.oftel.gov.uk*.

4. Lawrence Summers, "The New Wealth of Nations," speech to Hambrecht & Quist Technology Conference, 10 May 2000. At *www.treas.govt/pressreleases/ps617.htm*.

5. "Who Owns the Knowledge Economy?" *The Economist*, 8 April 2000.

6. Speaking at a technology forum at the University of California at Berkeley in May 2000. Quoted in "For Antitrust Cops, Microsoft Signals Need to Junk Old Models," by Alan Murray, *Wall Street Journal*, 9 June 2000.

7. While the basic point that the value of a network depends on the number of users is valid, it is only a rough guide. Other factors, such as the kind of inter-

· · · · ·

action among users, are also important and help to explain why, for instance, broadcast networks grow in different ways from point-to-point networks. For a fuller discussion, see Andrew Odlyzko, "The History of Communications and its Implications for the Internet." Unpublished preliminary manuscript at *www.research.att.com/~amo/doc.networks.html.*

8. Dan Roberts, "Watchdog Probes BT Over Portal Complaint," *Financial Times*, 17–18 June, 2000.

9. Bill Gates, "Compete, Don't Delete," *The Economist*, 13 June 1998.

10. Stan Liebowitz and Stephen Margolis, "The Fable of the Keys," *Journal of Law and Economics,* vol. 23, April 1990.

11. Stan Liebowitz and Stephen Margolis, "Winners, Losers, and Microsoft: Competition and Antitrust in High Technology" (Oakland, The Independent Institute, 1999).

12. Frances Cairncross, "The Death of Distance: A Survey of Global Telecommunications," *The Economist*, 30 September 1995, 26.

13. Atin Basu Choudhary, Robert Tollison, and William F. Shughart II, quoted in Richard B. Mackenzie and William F. Shughart II, "Why the Case for a Break-up Breaks Down," *Wall Street Journal*, 25 April 2000.

14. "Who Owns the Knowledge Economy?" *The Economist*, 8 April 2000.

15. "Patent Wars," *The Economist*, 8 April 2000.

16. James Gleick, "Patently Absurd," *New York Times*, 12 March 2000.

17. Quoted in "Patent Wars," *The Economist*.

18. Kevin Rivette and David Kline, *Rembrandts in the Attic* (Cambridge: Harvard Business School Press, 2000).

19. Thorold Barker and Dan Roberts, "The Cost of Forgetting Inventions," *Financial Times*, 24–25 June 2000.

20. "Patent Wars," *The Economist*.

21. James Gleick.

22. *Ibid*.

23. James Messen and Eric Maskin, "Sequential Innovation, Patents and Imitation," Working Paper 00-01, Department of Economics, Massachusetts Institute of Technology, January 2000.

24. "Who Owns the Knowledge Economy?" *The Economist*.

Chapter 11

1. Unofficial estimates suggest that two-thirds of Internet users in Saudi Arabia are women, many of them using it to avoid the restrictions on their business and social lives. "How Women Beat the Rules," *The Economist*, 2 October 1999.

2. Although more, of course, ultimately brings a high cost in terms of time. In particular, the fact that an extra e-mail costs nothing in financial terms increases its attractiveness to the sender, not to the recipient, and so ensures that the main increase will be in supply, not demand. Don Kluth, an eminent American computer scientist, has stopped using e-mail in order to ensure that anyone

who wants to contact him has first to overcome the extra barrier of using ordinary mail. Quoted in Andrew Odlyzko, "The History of Communications and its Implications for the Internet."

3. Survey by ITAC at *www.Telecommute.org*.

4. During BT's one-year experiment, eleven operators worked from home, monitored by psychologists from Aberdeen University, and handled 750,000 customer inquiries.

5. N. Ferguson, *The House of Rothschild: The World's Banker, 1849–1999* (New York Viking, 1999).

6. "Ask the Boiler," *The Economist*, 10 July 1999.

7. Work by Michael Morris, of Stanford Business School, and several colleagues finds that on-screen negotiations work best if conducted partly face-to-face, or between people who begin by swapping photographs or personal details, or who already know each other (*The Economist*, 8 April 2000).

8. Keith Naughton, "Cyberslacking," *Newsweek*, 29 November 1999, 62–65.

9. Charles Handy, *The Age of Unreason* (Boston: Harvard Business School Press, 1990).

10. Frances Cairncross, "A Connected World," *The Economist*, 13 September 1997. One partner kept a map in his filing cabinet to ensure that his virtual private office was set up identically each time he checked in.

11. Robert J. Saunders, Jeremy J. Warford, Björn Wellenius, *Telecommunications and Economic Development*, 2nd ed., International Bank for Reconstruction and Development (Baltimore, Md: Johns Hopkins Press, 1994), 130.

12. Study of four thousand adults in 2,700 households by Stanford Institute for the Quantitative Study of Society, 2000.

13. New York: Ace Books, 1984.

14. *World Telecommunication Development Report 1994* (Geneva: International Telecommunication Union, 1994), 14.

15. Stanford Institute for the Quantitative Study of Society.

16. Howard Rheingold, *The Virtual Community: Homesteading on the Electronic Frontier* (Reading, Mass: Addison-Wesley, 1993).

17. The idea was floated in *The Future of Work; Looking Ahead 2015* (London: Department of Trade and Industry, August 1999).

18. Thomas L. Friedman, *The Lexus and the Olive Tree* (New York: Farrar Straus & Giroux, 1999).

19. "Spanking Makes a Comeback," by Daniel Costelloa, *Wall Street Journal*, 9 June 2000.

20. Janice Maloney, "Volunteers Log On, Help Out," *New York Times*, 17 November 1999.

21. Daniel Miller and Don Slater, *The Internet: An Ethnographic Approach* (Oxford and New York: Berg, 2000).

22. "Search for Family Tree is Felled by Log-on Overload," *Financial Times*, 28 May 1999. Like so many new sites, it promptly crashed.

23. "After Powell," *The Economist*, 14 February 1998.

24. Quoted by David Crystal in *Cambridge Encyclopedia of the English Language* (Cambridge, England: Cambridge University Press, 1995), 386.

25. "Why the Pen is Mightier than the Mouse." Commencement address to students at the University of Southern California, republished in the *Financial Times*, 27–28 May 2000.

26. Michael Skapinker "The Tongue Twisters," *Financial Times*, 28 December 2000.

27. Crystal, *Cambridge Encyclopedia of the English Language*, 106.

28. Geoffrey Nunberg, quoted in "The Coming Global Tongue," *The Economist*, 21 December 1996.

29. Study by David Crystal, quoted in "The Coming Global Tongue."

30. Nunberg, quoted in "The Coming Global Tongue."

31. "In Japan, (*^o^*) Means Happiness," *International Herald Tribune*, 15 May 2000.

32. "Gumped," *The Economist*, 24 December 1994.

33. See *www.ntia.doc.gov*

34. *Information Technology Outlook* (Paris: OECD, 1997).

35. Victoria Griffiths, "Strategy for Block Release," *Financial Times*, 30 September 1996.

36. Pam Woodall, "Untangling the E-conomy," *The Economist*, 23 September 2000.

37. *Ibid.*

38. *Ibid.*

39. *On Your Farm*, produced by Alasdair Cross, 15 March 1998 BBC, Pebble Mill, Birmingham.

Notes: **bold** entries indicate more
substantial references; page references
followed by 'f' indicate that one or
more relevant figures are contained
within the page range shown.

About the Author

Frances Cairncross is a senior editor on the staff of *The Economist*, where she has worked since 1984. She has had responsibility for the Britain section and for coverage of the environment and, most recently, of media.

Cairncross is a graduate of Oxford University and Brown University. She is a governor of Britain's National Institute of Economic and Social Research; a member of the Council of the Institute for Fiscal Studies; an honorary fellow of St. Anne's College, Oxford; and a non-executive director of the Alliance & Leicester Group.

Her previous books include *Costing the Earth: The Challenge for Governments, the Opportunities for Business*, also published by Harvard Business School Press. She is married to Hamish McRae. They have two daughters and live in London.